SpringerBriefs in Statistics

More information about this series at http://www.springer.com/series/8921

Arilova A. Randrianasolo

Triple Double

Using Statistics to Settle NBA Debates

Arilova A. Randrianasolo
Department of Marketing, Lacy School of Business
Butler University
Indianapolis, IN, USA

ISSN 2191-544X   ISSN 2191-5458   (electronic)
SpringerBriefs in Statistics
ISBN 978-3-030-79031-8   ISBN 978-3-030-79032-5   (eBook)
https://doi.org/10.1007/978-3-030-79032-5

This Springer imprint is published by the registered company Springer Nature Switzerland AG
The registered company address is: Gewerbestrasse 11, 6330 Cham, Switzerland

Preface

Like millions of Americans, I look forward to tuning in to either ESPN or Fox Sports 1 multiple times per week to catch Stephen A. Smith on *First Take* lose his temper over Max Kellerman's sometimes "blasphemous" comments or watch Shannon Sharpe on FS1's *Undisputed* make his case to Skip Bayless on why Lebron James is better than Michael Jordan ever was. Typically, after watching a few debate points, I whip out my phone with conviction and send a message to a group text where a few friends and I debate our viewpoints on whatever NBA topic got me excited or even at times heated that day. These debates usually last all day long (thank god for unlimited text messages), and at the end of the day they go unresolved, usually concluding with lighthearted jabs and agreements to disagree, and every once in a while, someone in the group text would display their disagreements in all caps laced with emojis. Often, after a day of group text arguing, I am left with the question of how to truly settle these debates that seem to run in circles. I wrote this book to provide empirical evidence from statistical analyses on some of the most debated topics that NBA fans, reporters, players, and coaches have had to contemplate. Therefore, the goal of this book is to not only provide empirical evidence to clarify these debates for sports fans to use in their group texts but also bring practical implications to general managers and coaches across the NBA seeking to win championships as wells as journalists and sports analysts to reference. The debates I find most relevant to the NBA in current times are investigated throughout 7 chapters, as discussed in the following paragraphs.

Chapter 1 discusses who da real MVP is. Every single season, NBA fans fight over who they believe deserves the MVP award. For the weeks leading up to the announcement of each season's MVP, fans, journalists, and sports commentators weigh in on who deserves the most highly regarded individual award in the NBA. The MVP award is decided by votes casted by sports broadcasters and basketball analysts in the NBA world. Some would argue that this process is more of a popularity contest than an award for the actual most valuable player in the league. Therefore, this chapter breaks down what constitutes value for this award, which statistics contribute to building this value, and offers a formula on how to calculate da real MVP. The implications of this chapter reach more than just sports fans seeking to settle age-long debates on who deserves or deserved the MVP award, it also provides a perspective on how recruiters and general managers in the NBA can constitute value for their current and prospective players.

Chapter 2 takes on the task of naming the GOAT. So often, the term GOAT has been diluted to mean any player that has had some consistency in playing at a high level. Personally, I have heard that Lebron James, Kobe Bryant, Wilt Chamberlain, Michael Jordan, Bill Russell, Kareem Abdul-Jabbar, Magic Johnson, Shaquille O'Neal, and Tim Duncan, among many others, are all goats. This is a tribe of goats, when the term is meant to crown one player as the greatest of all time. With so many players that span across so many different eras, it is difficult to pinpoint who the actual GOAT is. I mean, the way basketball was played in the 1990s with infamous tough guys like Charles Barkley and Dennis Rodman was so much more physical than the late 2010s when Steph Curry and Dame Lillard regularly launched 3-pointers from near half-court. So how can players be compared across such different eras of the NBA as the game evolves. This chapter seeks to provide an answer to this question by analyzing this herd of goats and their dominance within their respective eras to name who the real GOAT is.

Chapter 3 discusses the myth of the super-team. Within the past decade, there has been an ongoing discussion among NBA fans that "super-teams" are ruining the NBA, or that some championships are worth less because some superstars decided to team up and dominate the league. This chapter investigates the accuracy of this sentiment. Is the super-team a new phenomenon? Weren't the showtime Lakers of the 1980s filled with superstars during their run (i.e., Kareem Abdul-Jabbar, James Worthy, Magic Johnson)? Didn't Larry Bird have a team with eventual hall of famers such as Kevin McHale and Robert Parish on his team to help win 3 championships? Do super-teams actually exist or is this just a myth? This chapter seeks to answer these questions.

Chapter 4 covers how All-NBA selections are made, and whether they are fair and accurate. Where the all-star selection is influenced by the celebrity of certain players since fans and media vote for their favorite players to make the all-star team, All-NBA honors might need to be more methodologically sound since these selections can have long-lasting effects on players' careers. In other words, if All-Star selections are chosen with subjective voting methods, shouldn't All-NBA selections be made with more objective measures? This chapter analyzes the selection process for all-NBA teams.

Chapter 5 analyzes the effectiveness of the "small ball" method, popularized by the Golden State Warriors between 2015 and 2019. The Warriors' "Hampton 5" line-up (sometimes called the death lineup), consisting of Kevin Durant, Stephen Curry, Klay Thompson, Draymond Green, and Andre Iguodala, were virtually unbeatable, but the small ball method did not work with every team that implemented it. This chapter empirically investigates the conditions in which the small ball method works well in the NBA, and when it is ineffective. This chapter seeks to aid NBA coaches who intend to implement the small ball method.

Chapter 6 discusses the "clutch gene" phenomenon in the NBA. Many media members, fans, and coaches have asserted that certain players just have the clutch gene, meaning that they are able to perform under pressure better than other players. Is this true, or is a manifestation of perception? This chapter investigates whether this clutch gene may be real, and if so, what effect it has on teams.

Chapter 7 discusses whether offense or defense matters more in winning championships. For those of us who played organized sports, there is a chance that we have heard a coach say, "offense wins games and defense wins championships." In the current era, when teams score upwards of 120 points per game, does defense really matter in winning championships more than offense? This investigates the answer to this question to provide an empirically based perspective on whether defensive or offensive oriented teams are more suited to win trophies.

Chapter 8 discusses the strategic implications of each of the previous 7 chapters with regard to how the results may be utilized by coaches, general managers, analysts, and media members to gain a better understanding of how to construct and find success with their teams.

Finally, Chap. 9 discusses debates that future research and books could explore. These debates are all relevant to the current NB world, yet fall outside of the scope and realm of this current book.

Overall, within this book, I seek to provide empirically driven answers to some of the most debated issues among NBA fans and sportscasters. I acknowledge that the beauty of sports in general lies in the comfort that we can all root for our favorite players and franchises but uncovering the objective answers to these debates can not only help bring clarity to basketball fans but also aid NBA executives, team general managers, recruiters, and coaches make decisions that may lead to their success.

Indianapolis, IN, USA   Arilova A. Randrianasolo

Acknowledgments

First and foremost, I would like to thank my significant other, Jeni, for providing me with the courage and support I needed to write this book. It is with this love and belief that I was able to work on being the best version of myself. Second, I'd like to thank my family for their enduring support of anything I do. Finally, I'd like to give a shout out the fellas in my group text, especially Ricky Thrash, Jr. for debating and discussing the topics of this book with me.

Contents

About the Author

Arilova (Lova) Randrianasolo is an assistant professor of marketing at Butler University's Lacy School of Business. He holds a bachelor's in international business and entrepreneurship from Saint Louis University, a master's degree in music business from New York University, and a doctorate in international business and marketing from Saint Louis University. Prior to this position at Butler University, he was an assistant professor of marketing at John Carroll University's Boler School of Business. Dr. Randrianasolo's research has been published in outlets such as *Journal of Brand Management, Journal of Global Marketing, Journal of International Marketing, Journal of International Consumer Marketing, Journal of Consumer Marketing,* and *Madagascar Conservation & Development.* Furthermore, he has presented original research at international conferences such as the *Academy of International Business*'s annual conference, the *American Marketing Association's* annual conference, and the *Academy of Marketing Science's* annual conference. Finally, Dr. Randrianasolo has experience in holding talks and seminars in developing countries. For example, in July 2019, he presented a talk at the US Embassy in Antananarivo, Madagascar, as well as held a seminar focused on non-profit marketing for non-profit organizations in Antananarivo, Madagascar.

Chapter 1
Da Real MVP

1.1 The Most Valuable Player Award

The National Basketball Association's most highly regarded and prestigious individual award is the Most Valuable Player (MVP) award. Who could forget when Shaquille O'Neal won the award in the 1999–2000 season by averaging nearly 30 points and 13 rebounds per game, or when Kevin Durant accepted the award with tears in his eyes and told his mother that she was "da real MVP"? This award has solidified the legendary career statuses of every player that has received it, and every MVP that is eligible for the Naismith Memorial Basketball Hall of Fame has been inducted. As one of the most revered trophies in basketball however there are so many questions every year about whether the correct player is or has been given the award. Should Kobe Bryant have won it in the 2005–2006 season over Steve Nash? I mean, Kobe did lead the league in scoring that year with over 35 points per game and adding 5.3 rebounds, 4.5 assists, and 1.8 steals per game in that season. Some would say Kobe got snubbed. Should Lebron have won it over Derrick Rose in the 2010–2011 season? I mean, Lebron was the most unstoppable force in the NBA that season. Has not Lebron been the most valuable player for like 10 years straight? Was James Harden robbed in the 2018–2019 season? I mean, those step-back threes are lethal!

The reason that these questions linger for years within the minds of basketball fans around the world is because the NBA has not clearly defined what "being valuable" means, and thus crowning a "most valuable player" becomes a subjective and opinion-based election. So, what does constitute value? Is it points scored? Is it how far you take your team? Is it being the only superstar on a winning team? Is it averaging a triple double? How can the NBA fairly assess value if it does not clearly define what is valuable? This chapter argues that value in the NBA has to do with contribution to winning, and that the most objective, fair, and uncontroversial way to determine MVPs is to statistically prove how the candidates contribute to

© The Author(s), under exclusive license to Springer Nature
Switzerland AG 2021
A. A. Randrianasolo, *Triple Double*, SpringerBriefs in Statistics,
https://doi.org/10.1007/978-3-030-79032-5_1

winning games. The following sections analyze basketball statistics to figure out who is da real MVP.

1.2   Which Statistics Matter?

Understanding what constitutes player value in the NBA is crucial if the most valuable player in league should be awarded a trophy every year. So, although we have all heard the arguments of how a player deserves the award because of their scoring or because of their leadership ability, the only real value that is generally accepted is the value in winning; that is winning games in the regular season since the MVP award does not consider the playoffs. Thus, to determine player value, we must determine which player contributes the most to winning in the regular season. What then contributes to winning? To answer this question, the following paragraphs report an analysis to determine which statistics contribute most to winning in the NBA.

To assess what contributes to winning, I collected data for each NBA team between 2014 and 2019.[1] There are 30 teams in the NBA, and data for 5 seasons were collected, and therefore there was a total of 150 observations. The dependent variable, in this case, is winning percentage since we are trying to figure out what contributes to winning. In terms of the independent variables, offensive statistics collected included: team assists per game, team points per game, team field goal percentage, and team offensive rebounds per game. Defensive statistics collected were team defensive rebounds per game, team steals per game, and team blocks per game. Also, team turnovers per game and team fouls per game were collected as these may influence winning. The season and team minutes per game were collected as control variables.

A multiple regression analysis in SPSS 26 was conducted, and the results showed that even when controlling for the year (the season) and the team minutes per game, the factors that significantly and positively contributed to winning percentage were points per game ($\beta = 0.25$, $p < 0.05$), field goal percentage ($\beta = 0.46$, $p < 0.001$), defensive rebounds ($\beta = 0.39$, $p < 0.001$), and steals per game ($\beta = 0.25$, $p < 0.05$). Turnovers per game ($\beta = -0.34$, $p < 0.05$) significantly and negatively impacted winning percentage. This regression model had an R Square of 0.71 with $p < 0.001$, meaning that this model predicts 71% of the variance in winning percentage.[2] Table 1.1 displays the results of the regression analysis.

So, what all of this jargon means is that the statistics that matter most to winning games (at least between 2014 and 2019) are points per game, field goal percentage, defensive rebounds, steals, and turnovers (low turnovers are better). So, with this in

---

[1] All statistics were collected from Basketball Reference (https://www.basketball-reference.com/).
[2] An R Square of 0.71 is pretty high, there is no model that will predict 100% of the variance in anything.

Table 1.1 Results of multiple regression analysis predicting winning percentage ($N = 150$)

	β	Standard error
Control variables		
Team minutes per game	0.001	0.05
Season (year)	−0.44***	0.05
Independent variables		
Assists per game	0.11	0.07
Points per game	0.25*	0.12
Field goal percentage	0.46***	0.08
Free throws percentage	0.05	0.05
Offensive rebounds per game	0.09	0.06
Defensive rebounds per game	0.39***	0.08
Steals per game	0.30***	0.06
Blocks per game	0.07	0.06
Turnovers per game	−0.34***	0.06
Fouls per game	−0.05	0.06
R Square: 0.71*		

*$p < 0.05$, **$p < 0.01$, ***$p < 0.001$

mind, has the NBA awarded the correct MVP for the seasons between 2014 and 2019? Let us look at each of the categories and determine if the NBA has gotten it right or wrong.

1.3   Scoring

There is nothing like watching a good shooter/scorer in an NBA game. I remember the night Kobe Bryant scored 81 points in 2006, or when Klay Thompson dropped 37 points in a single quarter in 2015. It is magical to watch players score at ease against some of the game's strongest defenders, and it is easy to see why points per game is always the first thing any sports announcer mentions when discussing MVP candidates. Yes, for sure, points matter, I mean the higher score wins the game, but according to the results of the regression equation, points per game must be considered along with field goal percentage. In fact, field goal percentage is a stronger predictor of winning percentage than points per game. Therefore, although it was impressive to watch Russell Westbrook score 31.6 points per game in his 2016–2017 season, we must keep in mind that he led the league in field goal attempts that year with 24 shots per game. That same year however James Harden fired off 18.9 shots per game, 5 shots less than Westbrook, and still managed to drop 29.1 points per game. Similarly, Isaiah Thomas launched 19.4 shots and managed to drop 28.9 points per game, and Anthony Davis took 20.3 shots and still had 28 points per game. So, was Westbrook's 31.6 points per game really more impressive than a player like James Harden's 29.1 points per game when Westbrook took more than 5

Table 1.2 The field goal percentage rankings of MVPs 2004–2019

Season	MVP	FG% ranking
2004–2005	Steve Nash	21
2005–2006	Steve Nash	17
2006–2007	Dirk Nowitzki	25
2007–2008	Kobe Bryant	68
2008–2009	Lebron James	34
2009–2010	Lebron James	26
2010–2011	Derrick Rose	83
2011–2012	Lebron James	12
2012–2013	Lebron James	5
2013–2014	Kevin Durant	27
2014–2015	Stephen Curry	28
2015–2016	Stephen Curry	26
2016–2017	Russell Westbrook	100
2017–2018	James Harden	81
2018–2019	Giannis Antetokounmpo	11

shots more than Harden each game? In fact, Russell Westbrook ranked 100th in the league in field goal percentage during his MVP year. Should the MVP award go to a player that had 99 players shoot the ball more efficiently? Table 1.2 displays the field goal percentage rankings of the MVPs from 2004 to 2019.

As shown in Table 1.2, in the seasons ranging from 2004 to 2019, the MVP winner has only ranked top 10 in field goal percentage once, and only six times in the top 25. If contributing to winning games makes a player valuable, then the NBA has failed to consider an important statistic, field goal percentage, when naming the league's most valuable player.

1.4 Defensive Rebounds

Rebounding is an essential part of any success in basketball. Nicknamed as "the big fundamental," the great Tim Duncan, heralded by many as one of the greatest power forwards to ever grace the court, averaged 10.8 rebounds throughout his decorated NBA career. Similarly, Kevin Garnett averaged 10 rebounds per game in his career. In fact, KG led the league in rebounds for multiple years, yet only won the MVP trophy once. No doubt, the NBA community recognizes rebounds as an important statistic when considering MVP candidates. However, is it really considered enough? And is the right type of rebounding considered? The result of the regression equation from earlier in this chapter suggests that defensive rebounds are what significantly contribute to winning games in the league, and in this specificity, the NBA seems to lack consideration. Table 1.3 displays the rankings of MVPs in defensive rebounds between 2004 and 2019.

Table 1.3 Defensive Rebound Rankings of MVPs between 2004 and 2019

Season	MVP	Defensive rebound ranking
2004–2005	Steve Nash	116
2005–2006	Steve Nash	69
2006–2007	Dirk Nowitzki	14
2007–2008	Kobe Bryant	40
2008–2009	Lebron James	14
2009–2010	Lebron James	19
2010–2011	Derrick Rose	101
2011–2012	Lebron James	18
2012–2013	Lebron James	15
2013–2014	Kevin Durant	16
2014–2015	Stephen Curry	104
2015–2016	Stephen Curry	60
2016–2017	Russell Westbrook	5
2017–2018	James Harden	46
2018–2019	Giannis Antetokounmpo	2

As shown in Table 1.3, in the 15 seasons between 2004 and 2019, the MVP has ranked top 10 in defensive rebounding only twice. It may not be as exciting to watch Andre Drummond grab a defensive rebound as it is to watch James Harden break a defender's ankles and hit a step-back three-pointer, but this does not mean that that defensive rebound is insignificant. The NBA community should place more emphasis on recognizing rebounders in the MVP races each year.

1.5 Steals

Offense wins games, defense wins championships; but does defense win MVPs? It is interesting that steals contribute to winning games since this statistic is rarely ever discussed in debates for who deserves the MVP trophy. The problem with this statistic is that players rarely get more than 3 steals per game, and the variance in steals across the league is extremely small. Specifically, the difference between the 2018–2019 season leader in steals (Paul George with 2.2 per game) and the player that came in 100th (Ryan Arcidiacono with 0.8 per game) is just 1.4 steals per game, but small incremental changes in steals can be significant contributors to winning games. This makes it so that many of those who may observe this statistic may not see a major difference between 1.2 steals per game and 2.0 steals per game, yet this 0.8 difference significantly impacts winning percentages. So, in terms of considering steals, has the NBA gotten it right with respect to choosing MVP winners? Table 1.4 displays the rankings of steals per game for MVPs between 2004 and 2019.

As shown in Table 1.4, the NBA may not have gotten this one wrong. Steph Curry won the MVP in the 2014–2015 season and ranked fourth in steals, and 1 year

Table 1.4 Steals rankings of MVPs between 2004 and 2019

Season	MVP	Steals ranking
2004–2005	Steve Nash	49
2005–2006	Steve Nash	78
2006–2007	Dirk Nowitzki	104
2007–2008	Kobe Bryant	9
2008–2009	Lebron James	8
2009–2010	Lebron James	11
2010–2011	Derrick Rose	52
2011–2012	Lebron James	3
2012–2013	Lebron James	12
2013–2014	Kevin Durant	32
2014–2015	Stephen Curry	4
2015–2016	Stephen Curry	1
2016–2017	Russell Westbrook	11
2017–2018	James Harden	5
2018–2019	Giannis Antetokounmpo	26

later, he took the trophy again while leading the league in steals per game. In fact, out of the 15 seasons displayed, the MVP has been top 25 in the league in steals per game 9 times, and top 50 in the league 12 times. Although I believe steals need to be given more consideration in the MVP votes, the MVPs have not been significantly low on number of steals.

1.6 Turnovers

As previously mentioned, in June 2017, Russell Westbrook was crowned the NBA's most valuable player (MVP) by averaging 31.6 points, 10.7 rebounds, and 10.4 assists per game in the 2016–2017 season. Although these statistics are impressive, in the same season, Westbrook also averaged 5.4 turnovers per game, which ranked him second in the league for turnovers per game in that season. James Harden, who was an MVP nominee that same year, was ranked number one in turnovers with 5.7 per game. In fact, Harden lead the league with 464 turnovers in the 2016–2017 season, while Russell Westbrook was second with 438 turnovers. The following year, James Harden won the MVP award and came in third with turnovers per game, while Russell Westbrook again came in second.

Undoubtedly, Westbrook's feat of averaging a triple double in his MVP season should be celebrated however as much as those triple doubles enforce his value, the number of turnovers should decrease it. In the current era of the NBA, MVP winners seem to turn the ball over more than in previous eras. In fact, in the 15 seasons between 2004 and 2019, there have been 12 MVP winners who have ranked top

10 in the league for turnovers, 5 of which ranked in the top 5. You might be thinking that the turnovers are higher just because MVP-level players usually handle the ball a lot more than other players. Well, in the 15 seasons prior to the 2004–2019 era (1989–2004), only six MVPs ranked top 10 in turnovers, and only 2 of which ranked top 5. Table 1.5 displays the MVP turnover rankings from each of the two eras.

To demonstrate the statistical difference between the two eras of the NBA shown in Table 1.5, an analysis of variance was conducted in SPSS 26 to compare the turnover rankings of MVPs in the two eras. There was a statistically significant difference in MVP turnover ranking between the 1989–2004 era and the 2004–2019 era at the $p < 0.05$ level [$F(1, 28) = 5.54$, $p = 0.026$]. What these results mean is that somewhere along the way, turnovers stopped mattering with respect to MVP consideration in the NBA.

Table 1.5 MVP turnover rankings for seasons 1989–2019

Season	MVP	Turnover ranking	Season	MVP	Turnover ranking
1989–1990	Magic Johnson	3	2004–2005	Steve Nash	8
1990–1991	Michael Jordan	37	2005–2006	Steve Nash	3
1991–1992	Michael Jordan	34	2006–2007	Dirk Nowitzki	53
1992–1993	Charles Barkley	10	2007–2008	Kobe Bryant	7
1993–1994	Hakeem Olajuwon	6	2008–2009	Lebron James	12
1994–1995	David Robinson	26	2009–2010	Lebron James	4
1995–1996	Michael Jordan	42	2010–2011	Derrick Rose	6
1996–1997	Karl Malone	28	2011–2012	Lebron James	7
1997–1998	Michael Jordan	45	2012–2013	Lebron James	11
1998–1999	Karl Malone	5	2013–2014	Kevin Durant	6
1999–2000	Shaquille O'Neal	23	2014–2015	Stephen Curry	10
2000–2001	Allen Iverson	6	2015–2016	Stephen Curry	8
2001–2002	Tim Duncan	10	2016–2017	Russell Westbrook	2
2002–2003	Tim Duncan	14	2017–2018	James Harden	3
2003–2004	Kevin Garnett	29	2018–2019	Giannis Antetokounmpo	5

1.7   So Who Is Da Real MVP?

So far in this chapter, I have discussed the statistics that should matter when considering MVP candidates in the NBA: points per game, field goal percentage, defensive rebounds, steals, and turnovers. The result of regression analysis showed that these statistics significantly contribute to winning games in the NBA's regular season. The beta for each significant predictor thus can be used to create a formula that calculates a player's value score:

$$
\begin{aligned}
\text{Value score} = {} & 0.25\left(\text{points per game}\right) + 0.46\left(\text{field goal percentage}\right) \\
& + 0.39\left(\text{defensive rebounds per game}\right) + 0.30\left(\text{steals per game}\right) \\
& - 0.34\left(\text{turnovers per game}\right)
\end{aligned}
$$

To utilize this formula, I collected data for players in the NBA in 2019–2020 on points per game, field goal percentage, defensive rebounds per game, steals per game, and turnovers per game. The statistics were standardized (with z scores) so they could all be in like terms. Each statistic was then plugged into the value score formula for each player and a value score was calculated. Table 1.6 displays the players with the top 10 value scores for the NBA's 2019–2020 season.

You might be wondering how Rudy Gobert could possibly be more valuable than the Greek Freak, Giannis Antetokounmpo. After all, in that MVP year, Giannis had an amazing season with 27.7 points, 12.5 rebounds, and 1.3 steals per game. Well, although Giannis' 27.7 points per game almost double Rudy's 15.9 points, Rudy led the league in field goal percentage at 66.9%, where Giannis was at 57.8%. In the value score formula, field goal percentage weighs more than points per game. Furthermore, although Giannis had more steals per game (1.3 to Rudy's 0.8), Rudy had far less turnovers than Giannis (1.6 to Giannis' 3.7). In the end, Giannis had an amazing season, but big Rudy, you da real MVP.

Table 1.6   2018–2019 top 10 MVP scores

MVP score	Player name
2.54	Rudy Gobert
2.25	Clint Capela
2.21	Andre Drummond
2.12	Giannis Antetokounmpo
1.84	DeAndre Jordan
1.66	Kawhi Leonard
1.57	Nikola Vučević
1.50	Steven Adams
1.50	Deandre Ayton
1.44	Paul George

Table 1.7 Top 10 MVP scores (2015–2018)

2017–2018 Season		2016–2017 Season		2015–2016 Season		2014–2015 Season	
MVP Score	Player name	MVP Score	Player name	MVP Score	Player name	MVP Score	Player name
2.59	DeAndre Jordan	3.13	DeAndre Jordan	3.12	DeAndre Jordan	3.69	DeAndre Jordan
2.55	Anthony Davis	2.33	Rudy Gobert	2.08	Andre Drummond	2.63	Anthony Davis
2.41	Clint Capela	2.06	Dwight Howard	1.92	Dwight Howard	2.36	Tyson Chandler
2.25	Andre Drummond	2.00	Anthony Davis	1.77	Hassan Whiteside	1.81	Kawhi Leonard
2.11	Karl-Anthony Towns	1.96	Andre Drummond	1.70	Anthony Davis	1.54	Rudy Gobert
1.98	Giannis Antetokounmpo	1.84	Hassan Whiteside	1.68	Kawhi Leonard	1.49	Andre Drummond
1.63	Steven Adams	1.70	Kevin Durant	1.33	Stephen Curry	1.43	Marcin Gortat
1.60	Larry Nance Jr.	1.50	Karl-Anthony Towns	1.28	Thaddeus Young	1.42	Nikola Vučević
1.50	Enes Kanter	1.46	Giannis Antetokounmpo	1.21	Karl-Anthony Towns	1.28	LaMarcus Aldridge
1.49	Otto Porter	1.45	Otto Porter	1.20	Marcin Gortat	1.27	Al Horford

1.8 Conclusion

Every year, before the NBA MVP winner is announced, basketball fans from all walks of life debate with their friends at bars, on social media platforms, in break rooms at work, on college campus dorm rooms, and at parties or social gatherings about why they think their favorite player deserves Maurice Podoloff the trophy. Skip Bayless goes on his sports debate show on Fox Sports One with Shannon Sharpe to make his case about who should win MVP, Stephen A. Smith and Max Kellerman debate on why someone was snubbed, and countless articles are written on all sorts of platforms discussing who really deserves the trophy. During all of this hoopla, everyone really knows that the MVP honors has turned into a popularity contest rather than an award for the player who holds the most value in the NBA. If this were not true, then how can a player come second in the league in turnovers and still win the MVP award?

If the current system really constituted the actual most valuable player, then perhaps in the past 10 seasons (2011–2020), more MVPs would have been champions in their MVP year. In the 2010s decade, the MVP has only gone on to win the championship 3 out of 10 times: Lebron James in 2012 and 2013 as well as Stephen Curry in 2015. Perhaps this also opens up the debate on whether MVP should

include the playoffs as well rather than just the regular season.[3] It has been repeat-
edly shown that players perform better in the playoffs than they do in the regular
season, I mean, after all, playoff games count for a lot more than regular season
games, and if the player is not in the playoffs, should they be considered in the
"most valuable" category anyway? No matter if the MVP should be considered
solely in the regular season or include the playoffs, one thing is for sure, the NBA
should be more objective in selecting who gets to hold that trophy and make
that speech.

---

[3] This debate is further discussed in Chap. 9.

Chapter 2
A Tribe of Goats

2.1 The Meaning of G.O.A.T.

The first time I heard the term G.O.A.T. was in the year 2000, when LL Cool J released his album by that very title, claiming he was the greatest (rapper) of all time. The self-proclaimed goat sparked controversy with his claim in the hip-hop world, igniting debates on who really was the goat. Although this term may not have originated with LL, since his album release, the term goat seemed to be used across all forums to debate the goats in all sorts of domains. The sports world was no different. Debates on Mike Tyson vs. Muhammad Ali, Lionel Messi vs. Cristiano Ronaldo, Pele vs. Maradona, and Lebron vs. Jordan have all sought to crown a goat.

Ask any basketball fan that lived through the 90s who the greatest of all time is and they will tell you that there is a reason why everyone wants to "be like Mike." Some would argue that there is a reason why when they crumble up a piece of paper and shoot it into a trashcan, they blurt out "Kobe!". An 18-year-old NBA fan could claim that Lebron is the greatest, I mean, after all, Lebron has been in the finals for 10 out of the 18 years of this fan's existence on earth. Still, a baby boomer could make the case for Kareem Abdul-Jabbar, as they may have witnessed Kareem score over 38,000 points in his career with arguably the most unstoppable shot ever: The Sky-hook. So, how can we settle once and for all who the goat is in basketball? To do this, we have to first discuss the definition of goat, then derive our research question from this definition. This chapter thus seeks to define the parameters of what makes someone a goat and empirically investigate who best fits this definition.

A. A. Randrianasolo, *Triple Double*, SpringerBriefs in Statistics,
https://doi.org/10.1007/978-3-030-79032-5_2

2.2 Defining the G.O.A.T.

To properly define the goat, it is important to understand the two components of this term: (1) greatest and (2) of all time. The second component is much easier to grasp than the first because "of all time" implies that the period examined spans the existence of the NBA. This means from 1946 until the current year. The first component of this term, "the greatest," needs more elaboration. If the "greatest" means the most championships, we can easily crown Bill Russell as the goat, and if the greatest means the most MVPs, then we can easily crown Kareem Abdul-Jabbar as the goat. However, all sports fans know that basketball is a team sport, and the parameters that allow great players to win championships and even MVPs can be dependent on the teams, coaches, and competition that the player faces each year. So, what then does it mean to be the greatest?

What makes this question so complicated is that the different players displayed their brilliance throughout different eras of the NBA. For example, guys like Bill Russell, Wilt Chamberlain, and Jerry West played before the NBA added a 3-point line on the court in 1979. So, their games were much different from Stephen Curry's game, where the baby-faced assassin could drop up to nine or ten 3-pointers in one game. Furthermore, many NBA analysts have stated that in the 90s players were much more physical than they are in today's game, and the game has evolved into much more of an outside game, rather than an inside game. So how do we define goat? Well, one thing remained the same throughout all the different eras of the NBA: the goal of each basketball game is for one team to beat the opposing team on the court. Thus, simply put, the goat is the player who has contributed the most to winning games for his team. Therefore, my definition of the greatest of all time here is: *the player who has done the most to help his team win throughout his career.*

The paragraphs in the following sections of this chapter describe how I calculated who the goat was with statistical analyses. It is important to note that these calculations represent a statistical perspective on the goat debate and that I am aware that there are many intangible factors that are not captured with these statistics. The pep talks in locker rooms to teammates, the tone set throughout the game, or handling different personalities on the team and coaching staff are all important factors that may contribute to wins, but the only objective components we have that is consistent across all eras is the player's and the teams' records of statistics.

2.3 Calculating the G.O.A.T Score

To calculate the G.O.AT. score, I started with the NBA's list of the greatest players of all time.[1] I took the top 20 players that ESPN list and collected data on each player's career. The players on this list are regarded by many as the greatest players of all time, and therefore it is worth analyzing whether their career stats really reflect their greatness in the goat debate. The ESPN list is shown in Table 2.1.

---

[1] https://www.espn.com/nba/story/_/id/29105801/ranking-top-74-nba-players-all-nos-10-1.

To commence the analyses, data was collected for teams that played in the NBA between 1956 and 2020.[2] 1956 was chosen as the starting point because this is the earliest season that a player on the top 20 list played. In other words, Bill Russell played in the earliest season from the players on the list, so his first season was the starting point. Statistics for each team in each season's free throw percentage, rebounds per game, assists per game, personal fouls per game, points per game, and win percentage for the regular season were collected. It is important to note that 3-point percentage, steals, blocks, and turnovers were not collected because (1) the NBA did not implement the 3-point line until 1979, and thus Bill Russell, Wilt Chamberlain, Jerry West, Oscar Robertson, Kareem Abdul-Jabbar, Julius Erving, and Moses Malone played a portion or the entirety of their careers without the 3-point option on offense, and (2) steals, blocks, and turnovers were not recorded in the NBA until 1973, so Bill Russell, Wilt Chamberlain, Jerry West, Oscar Robertson, Kareem Abdul-Jabbar, Julius Erving played a part or the entirety of their careers without recording steals. Data collection produced 1439 team observations. One observation is one team within one season, so, for example, the Los Angeles Lakers for the 2019–2020 season was one observation, while the Lakers of 2018–2019 was another.

Table 2.1   ESPN ranking for greatest players of all time

Rank	Player
1	Michael Jordan
2	Lebron James
3	Kareem Abdul-Jabbar
4	Bill Russell
5	Magic Johnson
6	Wilt Chamberlain
7	Larry Bird
8	Tim Duncan
9	Kobe Bryant
10	Shaquille O'Neal
11	Oscar Robertson
12	Hakeem Olajuwon
13	Stephen Curry
14	Kevin Durant
15	Julius Erving
16	Jerry West
17	Karl Malone
18	Moses Malone
19	Dirk Nowitzki
20	Kevin Garnett

[2]All statistics in this chapter were collected from www.basketball-reference.com.

After collecting the data, an analysis needed to be conducted to figure out what factors contributed to teams winning games within the era of 1956–2020. Therefore, a multiple regression analysis was then conducted in SPSS 26 to test the influence of free throw percentage, rebounds per game, assists per game, personal fouls per game, points per game on win percentage. The dependent variable here is win percentage for the regular season, and the other variables are the independent variables. The regression analysis yielded a significant model: F (5, 1435) = 83.46, $p < 0.001$, R Square = 0.23. The analysis results showed that assists per game (β = 0.30, $p < 0.001$), free throw percentage (β = 0.08, $p < 0.01$), rebounds per game (β = 0.18, $p < 0.001$), personal fouls (β = −0.36, $p < 0.001$), and points per game (β = 0.13, $p < 0.01$) all significantly contributed to winning percentage in the period between 1956 and 2020 in the NBA.

What these results show is that the strongest determinant of winning in the period examined was assists per game, while personal fouls had a strong and significant negative impact on winning. Overall, then, teams that had more assists, made a higher percentage of their free throws, rebounded better, scored more points, and had fewer personal fouls per game won more games in the era examined. Table 2.2 displays the results of the multiple regression analysis.

From the analysis results, a goat score can be calculated because it is shown how much each predictor contributes to the winning percentage of the team. The G.O.A.T. score formula derived from these results is:

$$\text{G.O.A.T score} = 0.30(\text{assists per game}) + 0.08(\text{free throw percentage})$$
$$+ 0.18(\text{rebounds per game}) - 0.36(\text{personal fouls}) + 0.13(\text{points per game})$$

After getting the goat score formula, I needed to collect the career statistics for each player to be able to plug their stats into the goat formula and see who the greatest is. Career stats on each of the players' free throw percentage, rebounds per game, assists per game, personal fouls per game, and points per game were thus collected. Table 2.3 displays the statistics for each player.

Table 2.2 Multiple regression results for the factors that influence winning percentage in the NBA between 1956 and 2020 ($N = 1439$)

	β	Standard error
Assists per game	0.30***	0.002
Free throw %	0.08**	0.13
Rebounds per game	0.18***	0.001
Personal fouls	−0.36***	0.002
Points per game	0.13**	0.001
R Square: 0.18***		

$^*p < 0.05$, $^{**}p < 0.01$, $^{***}p < 0.001$

Utilizing the statistics shown in Table 2.3 and the G.O.A.T score formula, a G.O.A.T. score was calculated for each of the 20 players. The scores for each player are displayed in Table 2.4. From this score, Wilt Chamberlain can be considered the greatest of all time, with his 30.1 points per game, 22.9 rebounds per game, 4.4 assists per game, 51% free throw percentage, and only 2 fouls per game. This finding is interesting because it puts Wilt Chamberlain at the top, while in the current sports climate, the debate always seems to be between Michael Jordan and Lebron James. What this tells us is that Wilt put up some monster numbers in the regular season.

This analysis was based on regular season statistics and took 64 seasons (1956–2020) into consideration. Some may argue that the G.O.A.T. score calculated here may not be representative of what really contributed to winning throughout different eras. For example, one basket may have contributed more to winning a

Table 2.3 Assists per game, free throw percentage, rebounds per game, personal fouls per game, and points per game for the 20 players in ESPN's list for greatest players of all time

Player	Seasons	Free throw %	Rebounds per game	Assists per game	Personal fouls per game	Points per game
Bill Russell	1956–1969	0.56	22.5	4.3	2.7	15.1
Wilt Chamberlain	1959–1973	0.51	22.9	4.4	2	30.1
Jerry West	1960–1974	0.81	5.8	6.7	2.6	27
Oscar Robertson	1960–1974	0.84	7.5	9.5	2.8	25.7
Kareem Abdul-Jabbar	1969–1989	0.72	11.2	3.6	3	24.6
Julius Erving	1971–1987	0.78	8.5	4.2	2.8	24.2
Moses Malone	1974–1995	0.76	12.3	1.3	2.4	20.3
Larry Bird	1979–1992	0.89	10	6.3	2.5	24.3
Magic Johnson	1979–1991	0.85	7.2	11.2	2.3	19.5
Hakeem Olajuwon	1984–2002	0.71	11.1	2.5	3.5	21.8
Michael Jordan	1984–1993; 1995–1998; 2001–2003	0.84	6.2	5.3	2.6	30.1
Karl Malone	1985–2004	0.66	10.1	3.6	3.1	25
Shaquille O'Neal	1992–2011	0.53	10.9	2.5	3.4	23.7
Kevin Garnett	1995–2016	0.79	10	3.7	2.4	17.8
Kobe Bryant	1996–2016	0.84	5.2	4.7	2.5	25
Tim Duncan	1997–2016	0.70	10.8	3	2.4	19
Dirk Nowitzki	1998–2019	0.88	7.5	2.4	2.4	20.7
Lebron James	2003–Present	0.73	7.4	7.4	1.8	27
Kevin Durant	2007–Present	0.88	7.1	4.1	1.9	27
Stephen Curry	2009–Present	0.91	4.5	6.6	2.5	23.6

Table 2.4 G.O.A.T. score for ESPN's top 20 players of all time

Player	Rank	G.O.A.T score
Wilt Chamberlain	1	8.68
Oscar Robertson	2	6.60
Lebron James	3	6.47
Magic Johnson	4	6.43
Bill Russell	5	6.38
Larry Bird	6	6.02
Michael Jordan	7	5.75
Jerry West	8	5.69
Kevin Durant	9	5.40
Kareem Abdul-Jabbar	10	5.27
Karl Malone	11	5.08
Stephen Curry	12	5.03
Julius Erving	13	4.99
Kobe Bryant	14	4.76
Shaquille O'Neal	15	4.61
Tim Duncan	16	4.51
Moses Malone	17	4.45
Kevin Garnett	18	4.42
Hakeem Olajuwon	19	4.38
Dirk Nowitzki	20	3.97

game in 1960 than it did in 2020 since 3 pointers did not exist in 1960. Therefore, the different eras of the NBA may need to be considered when discussing the G.O.A.T. This is discussed in the following section.

2.4   Adjusting the G.O.A.T Score for the Different Eras

To adjust for the different eras of the NBA, the dataset of 1439 was divided into 3 eras. The first era spanned from the 1956–1957 season until the 1978–1979 season since the 1979–1980 season was the first season with a 3-point line. The game was much different before the 3-point line, as the inside game had to be much more developed. Guys like Wilt Chamberlain and Bill Russell averaged over 22 rebounds per game in their careers during this era. This era includes 305 team observations. As was done in the first analysis, a multiple regression analysis was then conducted in SPSS 22 to test the influence of free throw percentage, rebounds per game, assists per game, personal fouls per game, and points per game on win percentage. The regression analysis yielded a significant model: F (6, 299) = 26.76, $p < 0.001$, R Square = 0.31. The analysis showed that assists per game ($\beta = 0.26$, $p < 0.001$), free throw percentage ($\beta = 0.17$, $p < 0.01$), rebounds per game ($\beta = 0.27$, $p < 0.001$), personal fouls ($\beta = -0.35$, $p < 0.001$), and points per game ($\beta = 0.22$, $p < 0.01$) all significantly contributed to winning percentage in the period between 1956 and 1979 in the NBA. It is interesting to note that the results reflect the conditions of the game in this era. When compared to the first analysis, this analysis showed results that reflect a higher importance of free throw shooting, rebounding, and points per game, while the influence of assists is slightly lower.

Table 2.5 displays the results of the multiple regression analysis. Using these results, a G.O.A.T. score for the first era of the NBA was constructed:

$$\text{First era G.O.A.T score} = 0.26\,(\text{assists per game}) + 0.17\,(\text{free throw percentage})$$
$$+ 0.27\,(\text{rebounds per game}) - 0.35\,(\text{personal fouls}) + 0.22\,(\text{points per game})$$

The second era of the NBA is between the 1979–1980 season and the 1997–1998 season. This era includes players like Larry Bird, Magic Johnson, Hakeem Olajuwon, Michael Jordan, and Karl Malone in their primes. This era included the dream team, regarded by many as the greatest era in NBA history. This era also had the introduction of the 3-point line, and so 3-point percentage was inserted as a predictor variable for this era. There were 480 team observations in this era. As was done in the previous analyses, a multiple regression analysis was then conducted in SPSS 22 to test the influence of free throw percentage, rebounds per game, assists per game, personal fouls per game, 3-point percentage, and points per game on win percentage. The regression analysis yielded a significant model: $F_{(6,\ 473)} = 41.59$, $p < 0.001$, R Square = 0.35. The analysis showed that assists per game ($\beta = 0.31$, $p < 0.001$), rebounds per game ($\beta = 0.21$, $p < 0.01$), personal fouls ($\beta = -0.25$, $p < 0.001$), 3-point percentage ($\beta = 0.24$, $p < 0.001$), and points per game ($\beta = 0.16$, $p < 0.05$) all significantly contributed to winning percentage in the period between 1979and 1998 in the NBA. It is important to note that in this era, free throw percentage was not a significant contributor to winning percentage. Table 2.6 displays the results of the multiple regression analysis. Using these results, a G.O.A.T. score for the first era of the NBA was constructed:

$$\text{Second era G.O.A.T score} = 0.31\,(\text{assists per game}) + 0.21\,(\text{rebounds per game})$$
$$- 0.26\,(\text{personal fouls}) + 0.24\,(3 - \text{point percentage}) + 0.16\,(\text{points per game})$$

The third era of the NBA is between the 1998–1999 season and the 2019–2020 season. This era includes players like Shaquille O'Neal, Kobe Bryant, Kevin Durant, Kevin Garnett, Dirk Nowitzki, Tim Duncan, and Stephen Curry in their primes. This

Table 2.5 Multiple regression results for the factors that influence winning percentage in the NBA between 1956 and 1979 ($N = 305$)

	β	Standard error
Assists per game	0.26***	0.004
Free throw %	0.17**	0.24
Rebounds per game	0.27***	0.001
Personal fouls	−0.35***	0.004
Points per game	0.22**	0.002
R Square: 0.31***		

$^*p < 0.05$, $^{**}p < 0.01$, $^{***}p < 0.001$

Table 2.6 Multiple regression results for the factors that influence winning percentage in the NBA between 1979 and 1998 ($N = 480$)

	β	Standard error
Assists per game	0.31***	0.004
Free throw %	0.03	0.249
Rebounds per game	0.21**	0.003
Personal fouls	−0.26***	0.003
3-point percentage	0.24***	0.121
Points per game	0.16*	0.002
R Square: 0.35* **		

*$p < 0.05$, **$p < 0.01$, ***$p < 0.001$

era included 654 team observations. Again, a multiple regression analysis was then conducted in SPSS 22 to test the influence of free throw percentage, rebounds per game, assists per game, personal fouls per game, 3-point percentage, and points per game on win percentage. The regression analysis yielded a significant model: F (6, 647) = 24.48, $p < 0.001$, R Square = 0.19. The analysis showed that assists per game ($\beta = 0.20$, $p < 0.001$), rebounds per game ($\beta = 0.11$, $p < 0.05$), 3-point percentage ($\beta = -0.22$, $p < 0.001$), points per game ($\beta = 0.27$, $p < 0.001$), and personal fouls ($\beta = -0.19$, $p < 0.001$) all significantly contributed to winning percentage in the period between 1998 and 2020 in the NBA. It is important to note that in this era, free throw percentage was not a significant contributor to winning percentage, similar to the second era. Also, interesting to note is that 3-point percentage negatively influenced winning percentage. This is interesting because in the modern NBA, 3-point shooting seems to be a necessity to win games and championships. Superstars like Stephen Curry and Klay Thompson built their careers as the Splash Brothers off of shooting from beyond the arc. Similarly, James Harden won an MVP off of his signature step back 3-point shot. So how can 3-point shooting negatively impact winning? Well, this may be a sign that teams in the current era are overloading on 3-pointers. A lot of teams (i.e., 2019 Houston Rockets) have built squads focused on outside shooting, and perhaps the league has hit the point where there is too much 3-point shooting. What worked for Stephen Curry may not work for every team. Table 2.7 displays the results of the multiple regression analysis. Using these results, a G.O.A.T. score for the first era of the NBA was constructed:

$$\text{Second era G.O.A.T score} = 0.20(\text{assists per game}) + 0.11(\text{rebounds per game})$$
$$- 0.19(\text{personal fouls}) - 0.22(3-\text{point percentage}) + 0.27(\text{points per game})$$

Utilizing the goat score formula for each era on the players from ESPN's list that played in that era, a goat score was calculated for each of the 20 players. In other words, each player's goat score was calculated based on their era's goat score. Table 2.8 displays the goat list.

When the goat score is adjusted for the eras in the NBA, it is interesting to note that Wilt Chamberlain seems to have picked up more points in the goat score, while Michael Jordan fell on the list. But how can this be? How could his royal airness fall

Table 2.7 Multiple regression results for the factors that influence winning percentage in the NBA between 1998 and 2020 ($N = 654$)

	β	Standard error
Assists per game	0.20***	0.003
Free throw %	0.06	0.20
Rebounds per game	0.11*	0.003
Personal fouls	−0.19***	0.003
3-point percentage	−0.22***	0.004
Points per game	0.27**	0.002
R Square: 0.19*		

$^*p < 0.05$, $^{**}p < 0.01$, $^{***}p < 0.001$

Table 2.8 G.O.A.T. score adjusted for eras for ESPN's top 20 players of all time

Rank	Player	Era	GOAT score
1	Wilt Chamberlain	1	13.34
2	Bill Russell	1	9.67
3	Oscar Robertson	1	9.31
4	Lebron James	3	9.17
5	Jerry West	1	8.48
6	Kevin Durant	3	8.45
7	Kareem Abdul-Jabbar	1	8.44
8	Julius Erving	1	7.86
9	Kobe Bryant	3	7.71
10	Stephen Curry	3	7.62
11	Magic Johnson	2	7.58
12	Shaquille O'Neal	3	7.44
13	Moses Malone	1	7.41
14	Larry Bird	2	7.38
15	Michael Jordan	2	7.16
16	Karl Malone	2	6.50
17	Tim Duncan	3	6.42
18	Dirk Nowitzki	3	6.35
19	Kevin Garnett	3	6.13
20	Hakeem Olajuwon	2	5.73

so far down the list? One answer may be that the Jumpan really took off in the play-offs rather than the regular season, and these goat scores are calculated based on players' regular season statistics. In other words, thus far, I have displayed the results of the goat score while taking 64 years of NBA seasons into consideration as well as the goat score if the formula is adjusted for each era however these statistics only consider regular season statistics, and many NBA fans and sports personalities know that greatness shines in the playoffs. So, to account for this, the following section examines the goat formulas in terms of player playoff statistics rather than regular season statistics.

2.5   G.O.A.T in the Playoffs

In order to calculate the goat score for playoff statistics, data for the career playoff statistics were collected for each of the 20 players. These statistics are shown in Table 2.9.

Utilizing the players' playoff career statistics, the goat score was first calculated with the goat score formula that was not adjusted for different eras, and the results are shown in Table 2.10.

After the goat score was calculated for the playoff statistics with the overall NBA goat score formula, a goat score for each player was calculated with the era-adjusted goat score formulas. The results of these calculations are shown in Table 2.11.

Table 2.9 Career playoff statistics for 3-point percentage, assists per game, free throw percentage, rebounds per game, personal fouls per game, and points per game for the 20 players in ESPN's list for greatest players of all time

Player	3-point FG %	Free throw %	Rebounds per game	Assists per game	Personal fouls per game	Points per game
Bill Russell	N/A	0.60	24.9	4.7	3.3	16.2
Wilt Chamberlain	N/A	0.47	24.5	4.2	2.6	22.5
Jerry West	N/A	0.81	5.6	6.3	2.9	29.1
Oscar Robertson	N/A	0.86	6.7	8.9	3.1	22.2
Kareem Abdul-Jabbar	N/A	0.74	10.5	3.2	3.4	24.3
Julius Erving	N/A	0.78	8.5	4.4	2.9	24.2
Moses Malone	N/A	0.76	14	1.5	2.7	22.1
Larry Bird	0.32	0.89	10.3	6.5	2.8	23.8
Magic Johnson	0.24	0.84	7.7	12.3	2.8	19.5
Hakeem Olajuwon	0.22	0.72	11.2	3.2	3.9	25.9
Michael Jordan	0.33	0.83	6.4	5.7	3	33.4
Karl Malone	0.16	0.74	10.7	3.2	3.4	24.7
Shaquille O'Neal	N/A	0.50	11.6	2.7	3.6	24.3
Keving Garnett	0.27	0.79	10.7	3.3	3	18.2
Kobe Bryant	0.33	0.82	5.1	4.7	3	25.6
Tim Duncan	0.14	0.69	11.4	3	2.8	20.6
Dirk Nowitzki	0.37	0.89	10	2.5	2.8	25.3
Lebron James	0.34	0.74	9	7.2	2.3	28.8
Kevin Durant	0.35	0.86	7.7	4	2.5	29.1
Stephen Curry	0.40	0.91	5.4	6.3	2.5	26.5

Table 2.10 Playoff G.O.A.T. score for ESPN's top 20 players of all time

Rank	Playoff GOAT score	Player
1	7.70	Wilt Chamberlain
2	6.86	Bill Russell
3	6.76	Lebron James
4	6.67	Magic Johnson
5	6.19	Michael Jordan
6	5.96	Larry Bird
7	5.71	Oscar Robertson
8	5.70	Jerry West
9	5.54	Kevin Durant
10	5.48	Stephen Curry
11	5.01	Julius Erving
12	4.93	Moses Malone
13	4.93	Karl Malone
14	4.90	Dirk Nowitzki
15	4.84	Kareem Abdul-Jabbar
16	4.80	Shaquille O'Neal
17	4.68	Tim Duncan
18	4.64	Kobe Bryant
19	4.50	Hakeem Olajuwon
20	4.27	Kevin Garnett

Table 2.11 Playoff G.O.A.T. score adjusted for era for ESPN's top 20 players of all time

Rank	Playoff adjusted GOAT score	Player
1	11.83	Wilt Chamberlain
2	10.46	Bill Russell
3	9.55	Lebron James
4	8.81	Kevin Durant
5	8.67	Jerry West
6	8.31	Stephen Curry
7	8.22	Moses Malone
8	8.07	Oscar Robertson
9	7.95	Kareem Abdul-Jabbar
10	7.88	Julius Erving
11	7.82	Magic Johnson
12	7.68	Michael Jordan
13	7.68	Dirk Nowitzki
14	7.62	Kobe Bryant
15	7.47	Shaquille O'Neal
16	7.26	Larry Bird
17	6.66	Tim Duncan
18	6.47	Hakeem Olajuwon
19	6.31	Karl Malone
20	5.96	Kevin Garnett

2.6   Conclusion

Since he retired for the final time in 2003, virtually all sports media and personalities have heralded Michael Jordan as the undisputed G.O.A.T. 6 championships, 6
finals MVPs, 5 MVPs, 10 time scoring champion, the most famous logo (jumpman)
on basketball shoes, and unbelievable swagger, it is easy to make an argument for
Michael Jordan to be regarded as the best to ever do it. However, after every single
analysis and calculation I made in this chapter, Wilt Chamberlain topped the list
based on his career statistics. With 2 championships, 1 finals MVP, 4 MVPs, and 7
scoring titles, it may be easy to dismiss Wilt as being inferior to the man that made
the #23 famous. When we look at accolades, MJ is superior to Wilt however career
statistics may tell a different story. Table 2.12 compares the career statistics for
these two players.

As shown in the table, both players have 30.1 points per game, which means they
are both scoring machines, I mean, Wilt once dropped 100 points in a game. MJ has
Wilt beat in free throw percentage and assists per game, but Wilt has nearly 4 times
as many rebounds per game as MJ. For a player to score 30 points and grab 22
rebounds a game consistently is no easy feat, and these rebounds are what boosted
Wilt's goat score to be so much higher than MJ. Wilt is the G.O.A.T., at least according to career statistics.

It is important to note, once again, that this chapter (and this book) provides a
statistical breakdown of debates, in this case, the goat debate, and there may be
intangible or unrecorded components that contribute to greatness. So, if someone
were to tell me that MJ or Lebron is the greatest, I can respect their view, but one
thing is for sure, based on stats, Wilt is #1.

Table 2.12 Comparison of Wilt Chamberlain and Michael Jordan career statistics implication:
3-point shooting

	Points per game	Assists per game	Free throw percentage	Rebounds per game	personal fouls per game
Michael Jordan	30.1	5.3	83.5	6.2	2.6
Wilt Chamberlain	30.1	4.4	51.1	22.9	2

Chapter 3
The Myth of the Super-Team

3.1 What Is a Super-Team?

In the summer of 2010, the whole sports world was anxiously awaiting Lebron James' decision on which team he would sign with to continue his already remarkable career. Lebron had spent the previous 7 seasons as a Cleveland Cavalier, and as a Cav, he had reached the NBA finals once and earned 2 MVP trophies up to that point. On July 1, 2010, Lebron became an unrestricted free agent, and his future as a Cavalier was uncertain. He was 25 years old at the time and was already considered the undisputed best basketball player in the world by many players and sports personalities, and, thus, teams had spent considerable effort recruiting him to come join their squads. I was in New York City at the time, and I remember Knicks fans wholeheartedly believing that Lebron would be wearing blue and orange the following year, and also the anger when they found out that he was not. Well, on July 8, 2010, during an infamous ESPN special called "The Decision," King James announced to the world that he would be taking his talents to South Beach and sign with the Miami Heat. This decision meant that Lebron would team up with superstars Dwyane Wade and Chris Bosh as he entered his physical prime. Essentially, Lebron would be forming a Super-team with 2 other all-stars.

After Lebron's announcement, NBA fans in Cleveland burned and stomped his jersey in the streets, and Lebron was stated by many to just be "ring-chasing" and to be "taking the easy way out." These sentiments always baffled me for several reasons. First, these notions seem to hint that playing basketball at the highest level and wanting to win championships with other superstars is the "easy way out." Second, within the NBA's history, there have been tons of championship teams that have had multiple superstars on them, from the showtime Lakers to Tim Duncan's Spurs. The difference, it seemed, was that Lebron was making the choice to team up with other superstars and the team was not the result of general managers moving chess pieces around the league. I guess when players construct teams, it is the easy way out, but

A. A. Randrianasolo, *Triple Double*, SpringerBriefs in Statistics, https://doi.org/10.1007/978-3-030-79032-5_3

when general managers and owners work their magic to create all-star squads, it is the legitimate route to glory.

In the summer of 2016, Kevin Durant experienced something remarkably similar to what Lebron experienced 6 years prior. Durant's contract with the Oklahoma City Thunder, the team he had spent the previous 9 seasons with, had come to an end. During his tenure as a Thunder, Durant was able to gain an MVP award, 4 scoring titles, and was able to join the 50-40-90 club[1] in 2013. On July 4, 2016, Kevin Durant announced to the world that he would be joining the Golden State Warriors. Durant made this decision after his Thunder team that previous season had lost to the Warriors in the western conference finals. They lost this series after being up 3-1. The Warriors that year won a record 73 wins in the regular season, so Durant's decision was regarded by many as unrespectable because he was "taking the easy way out by joining a 73-win team" and "joining the very team that he couldn't beat." In fact, I remember watching ESPN's First Take show and seeing Stephen A. Smith state that Durant's decision was the weakest move he had ever seen in basketball history. Russell Westbrook, Durant's all-star teammate while he was a Thunder, tweeted a picture of a cupcake on the day Durant made his announcement, which many took as a subtle jab that called Durant soft. When Durant made his first trip back to OKC to face the Thunder in his Golden State jersey, countless fans in the stands had cupcake signs.

Why is there such backlash on the decisions that guys like Lebron or Kevin Durant make? Every off-season, tons of players switch teams, either through trade or through free agency signings. The difference is that (1) the player in question (e.g., Lebron James, Kevin Durant) is a major superstar and (2) the player in question decides to join other superstars. When superstar players do this, they create what has been stated by many to be a super-team. So, what is this super-team they speak of? It is defined by a sentiment of superstars not being able to win with a team they are on, so they somehow figure out a way to get other superstars to join them. From this sentiment, I define a *super-team as a team constructed by superstar players seeking to team up in order to be contenders for a championship*. This definition means that a super-team is not constructed by general managers, team owners, or coaches, but rather superstars who use their influence to get other stars to join them.

So, which teams are super-teams? To answer this question, I reviewed teams that made the NBA finals within the past 15 seasons (2005–2020) to determine super-teams based on the definition. The following teams can thus be considered super-teams:

- **2019–2020 Los Angeles Lakers**. *Justification*: This core of this team includes multiple-time all-stars, Lebron James and Anthony Davis. Anthony Davis joined the Lakers after Lebron let it be known publicly in the previous season that he was unhappy with the current team he had and needed AD to join him.

---

[1] The 50-40-90 club is for players who achieve an average of at least 50% field goal percentage, 40% 3-point field goal percentage, and 90% free throw percentage in one season.

- **2016–2019 Golden State Warriors**. *Justification*: Kevin Durant, a former MVP, joined this team in 2016 through free agency, which already included all-stars Stephen Curry, Klay Thompson, Andre Iguodala, and Draymond Green. This created the "Hampton 5" lineup, which is a result of Stephen Curry, Klay Thompson, Andre Iguodala, and Draymond Green traveling to the Hamptons to go recruit Kevin Durant.
- **2014–2018 Cleveland Cavaliers**. *Justification*: In 2014, all-stars Lebron James and Kevin Love joined all-star point guard Kyrie Irving through free agency. In fact, I remember some reports that Kevin Love would not go to the Cavs unless he was paired with Lebron.
- **2010–2014 Miami Heat**. *Justification*: In 2010, all-stars Lebron James and Chris Bosh joined all-star and former finals MVP Dwyane Wade to construct one of the most notorious super-teams of all time.

In the last 15 seasons, the following teams that made the NBA finals are thus not considered super-teams:

- **2019–2020 Miami Heat**. *Justification*: At the time Jimmy Butler joined the Miami Heat, there were not other all-stars on the team. Bam Adebayo made the all-star team only after Butler joined the team.
- **2018–2019 Toronto Raptors**. *Justification*: Although Kawhi Leonard is a multiple-time all-star, he was traded to the Raptors by the Spurs and did not sign while he was a free agent. In fact, Kawhi went to the Los Angeles Clippers as soon as his contract was done with the Raptors, even after winning a championship.
- **2014–2015 Golden State Warriors**. *Justification*: During this period (before Kevin Durant), the all-stars on the team (Stephen Curry, Klay Thompson, Draymond Green) were all drafted by the Warriors organization.
- **2005–2014 San Antonio Spurs**. *Justification*: The all-stars of this team (i.e., Tony Parker, Manu Ginobili, and Tim Duncan) were all "home-grown," and were not a manifestation of players teaming up.
- **2011–2012 Oklahoma City Thunder**. *Justification*: All of the key players on this team (i.e., Kevin Durant, Russell Westbrook, James Harden, Serge Ibaka) were either drafted by the Thunder or were not all-stars when they joined the team.
- **2005–2011 Dallas Mavericks**. *Justification*: This team only had one superstar, Dirk Nowitzki, and he was "home-grown."
- **2008–2010 Los Angeles Lakers**. *Justification*: Although Pau Gasol was an all-star when he joined Kobe Bryant on the Lakers in 2008, he did not join in free agency, but rather through a trade that was manufactured by the Lakers organization.
- **2007–2010 Boston Celtics**. *Justification*: Both Kevin Garnett and Ray Allen were all-stars when they joined Paul Pierce on the Celtics, but they joined through trades.
- **2006–2007 Cleveland Cavaliers**. *Justification*: Lebron James was the only superstar on this team.

- **2005–2006 Miami Heat**. *Justification*: Although Shaquille O'Neal joined this team through free agency, Dwyane Wade was not yet an all-star when he joined.

After classifying teams that made the NBA finals in the past 15 seasons as either super-teams or not, the question that needs to be answered is: do super-teams ruin competition in the NBA? In other words, when superstars team up by their own power and will, does it create an unfair advantage as compared when superstars are brought together by general managers, coaches, and owners?

To answer this question, data was collected on the 30 teams that have made the NBA finals in the 15 seasons between 2005 and 2020 to be analyzed. The independent variable in this analysis is whether the team is a super-team or not. There were three dependent variables collected for this analysis, as described below:

1. **Regular season winning percentage**. The logic behind this dependent variable is that if super-teams really did destroy or diminish the competitive landscape of the NBA by creating unfair advantages, then there should be a significant difference between the regular season winning percentage of super-teams and non-super-teams.
2. **Number of playoff games played to get to the finals.** The logic here is that if super-teams really did destroy or diminish the competitive landscape of the NBA by creating unfair advantages, then it would take super-teams less games to get to the finals than non-super-teams. In other words, super-teams should have shorter series in the playoffs before the finals, and therefore less playoff games than non-super-teams if they have a competitive advantage.
3. **Number of games won in the finals**. The logic behind this dependent variable is that if super-teams in fact have a competitive advantage, then they should win more games in the NBA finals than non-super-teams. Table 3.1 displays the data for each of the 30 teams.[2]

To test the relationships between the independent variable and the dependent variables, 3 analyses of variances (ANOVA) in SPSS 26 were conducted to test the differences between regular season winning percentage, number of playoff games, and number of games won in the finals across the two groups (super-teams and non-super-teams). With respect to regular season winning percentages, there were no statistically significant differences between group means as determined by the one-way ANOVA ($F(1,29) = 0.481$, $p = 0.49$). With respect to the number of playoff games played to get to the finals, there were no statistically significant differences between group means as determined by the one-way ANOVA ($F(1,29) = 0.463$, $p = 0.50$). With respect to regular season winning percentages, there were no statistically significant differences between group means as determined by the one-way ANOVA ($F(1,29) = 0.290$, $p = 0.59$). These results are displayed in Table 3.2.

---

[2] All Statistics were collected from www.basketball-reference.com.

Table 3.1 Regular season winning percentage, number of games it took to get to the finals, number of games won in the finals for super-teams, and non-super-teams

Season	Team	Regular season winning percentage	# of games it took to get to the finals	# games won in finals	Super-team
2019–2020	Los Angeles Lakers	0.732	15	4	YES
2019–2020	Miami Heat	0.603	16	2	NO
2018–2019	Toronto Raptors	0.707	15	4	NO
2018–2019	Golden State Warriors	0.695	16	2	YES
2017–2018	Golden State Warriors	0.707	17	4	YES
2017–2018	Cleveland Cavaliers	0.610	17	0	YES
2016–2017	Golden State Warriors	0.817	12	4	YES
2016–2017	Cleveland Cavaliers	0.622	18	1	YES
2015–2016	Cleveland Cavaliers	0.695	14	4	YES
2015–2016	Golden State Warriors	0.890	17	3	NO
2014–2015	Golden State Warriors	0.817	15	4	NO
2014–2015	Cleveland Cavaliers	0.646	16	2	YES
2013–2014	San Antonio Spurs	0.756	18	4	NO
2013–2014	Miami Heat	0.659	19	1	YES
2012–2013	Miami Heat	0.805	16	4	YES
2012–2013	San Antonio Spurs	0.707	15	3	NO
2011–2012	Miami Heat	0.697	18	4	YES
2011–2012	OKC Thunder	0.712	14	1	NO
2010–2011	Dallas Mavericks	0.695	16	4	NO
2010–2011	Miami Heat	0.707	16	2	YES
2009–2010	Los Angeles Lakers	0.695	16	3	NO
2009–2010	Boston Celtics	0.610	17	4	NO

(continued)

Table 3.1 (continued)

Season	Team	Regular season winning percentage	# of games it took to get to the finals	# games won in finals	Super-team
2008–2009	Los Angeles Lakers	0.793	18	4	NO
2008–2009	Orlando Magic	0.756	19	1	NO
2007–2008	Boston Celtics	0.805	20	4	NO
2007–2008	Los Angeles Lakers	0.695	15	2	NO
2006–2007	San Antonio Spurs	0.707	16	4	NO
2006–2007	Cleveland Cavaliers	0.610	16	0	NO
2005–2006	Miami Heat	0.634	17	4	NO
2005–2006	Dallas Mavericks	0.732	19	2	NO

Table 3.2 ANOVA results for group differences between super-teams and non-super-teams with respect to regular season winning percentage, playoff games needed to get to the finals, and games won in the NBA finals between 2005 and 2020

	DV: winning percentage				
	SS	MS	df	F	p
Between groups	0.003	0.003	1	0.481	0.49
Within groups	0.146	0.005	28		
Total	0.148		29		
	DV: playoff games needed to get to finals				
	SS	MS	df	F	p
Between groups	1.422	1.422	1	0.463	0.50
Within groups	85.94	3.069	28		
Total	87.37		29		
	DV: number of games won in the finals				
	SS	MS	df	F	p
Between groups	0.556	0.556	1	0.29	0.59
Within groups	53.61	1.92	28		
Total	54.17		29		

3.2 Conclusion

The memes, sports debates, and discussions in the sports world about super-teams have been prevalent within the past decade, particularly enhanced by Lebron James' decision to take his talents to South Beach and Kevin Durant joining the Golden State Warriors' juggernaut team in 2016. Even recently in 2021, when James Harden requested to be traded from the Houston Rockets and joined Kevin Durant and Kyrie Irving in Brooklyn, the sentiment that he was taking the easy way to "ring-chase" because he could not get it done on his own in Houston was immediately discussed on sports forums. The results of the analysis in this chapter provide evidence that super-teams do not decrease the level of competition in the NBA, and that super-teams do not make it easier for anyone to win championships. Yes, when superstars team up on their own accord, they tend to have really good teams that are hard to beat, yet the same can be stated for teams where superstars organically develop and are "home-grown" on a team. When basketball players enter the league, they usually have no control over where they end up as they are drafted by teams. Many players get traded by teams without their input, yet when stars are able to control who they play with, it seems to be frowned upon. Players like Kevin Durant and Lebron James spent years to reach their superstar status, and they earn their accolades every moment they play on the court, so why should not they be able to decide where and with whom they play with? The super-team is a myth, and the truth is that championships are won with a team, and usually, championship teams have more than one star. So, why hate on stars who want to play and win with other stars?

Chapter 4
Hey Now, You are an All-Star…But are you All-NBA?

4.1 The All-NBA Selection

The All-NBA selection is a yearly selection of players who have shown to be the best at their positions throughout each season, and consists of three team selections: first team, second team, and third team All-NBA. For each team, two guards are chosen, two forwards are chosen, and one center is chosen. Therefore, 6 guards total make All-NBA each year, as well as 6 forwards, and 3 centers. These 15 players are supposed to represent the best in their particular position within the NBA. However, with the current system of voting, do the results actually reflect this? In this chapter, I present a case that the way in which the NBA selects All-NBA teams are unfair and should be changed.

The selections are determined with a vote from a panel of sports media members (e.g., broadcasters, journalists, TV personalities). In May 2019, the NBA announced the All-NBA selection for the 2018–2019 season. The list of that year is shown in Table 4.1.

From this selection, one notable shooting guard was missing: Klay Thompson. At the time of this All-NBA selection's announcement to the general public, Klay was preparing for his fifth trip to the finals alongside fellow Warriors Stephen Curry and Draymond Green. Hours after the announcement, Klay, who seemed like he had not had the chance to look at the list, was interviewed by reporters and asked how he felt about not making the All-NBA selection, particularly noting that guard Kemba Walker had made the third team and he had not. I remember immediately feeling confused and laughing out loud as Klay rolled his eyes and said "Whatever, I'd rather win a championship than be third team All-NBA." Klay's reaction was a highly debated topic on sports radio and sports debates shows for the next 24 h, with most sports journalists asking the question of whether a member of the Splash Brothers who had helped his team get to a fifth straight trip to the NBA finals had been snubbed.

© The Author(s), under exclusive license to Springer Nature
Switzerland AG 2021
A. A. Randrianasolo, *Triple Double*, SpringerBriefs in Statistics,
https://doi.org/10.1007/978-3-030-79032-5_4

Table 4.1 All-NBA selections for the 2018–2019 season

Position	First team	Second team	Third team
Guard	James Harden	Damian Lillard	Kemba Walker
Guard	Stephen Curry	Kyrie Irving	Russell Westbrook
Forward	Giannis Antetokounmpo	Kevin Durant	Lebron James
Forward	Paul George	Kawhi Leonard	Blake Griffin
Center	Nikola Jokic	Joel Embiid	Rudy Gobert

Table 4.2 Statistics for the 2018–2019 season for Klay Thompson and the other guards who made the All-NBA selections

Player	FG%	3P%	2P%	FT%	AST	STL	TOV	PTS
Klay Thompson	46.7	40.2	51.6	81.6	2.4	1.1	1.5	21.5
James Harden	44.2	36.8	52.8	87.9	7.5	2	5	36.1
Stephen Curry	47.2	43.7	52.5	91.6	5.2	1.3	2.8	27.3
Damian Lillard	44.4	36.9	49.9	91.2	6.9	1.1	2.7	25.8
Kyrie Irving	48.7	40.1	53.3	87.3	6.9	1.5	2.6	23.8
Russell Westbrook	42.8	29.0	48.1	65.6	10.7	1.9	4.5	22.9
Kemba Walker	43.4	35.6	49.4	84.4	5.9	1.2	2.6	25.6

Although Klay is right, championships are much better than a third team All-NBA selection, the "snub" impacted his career with regards to potential earnings. Specifically, if Klay Thompson had made the All-NBA team, he would have been eligible for a "supermax" contract worth $221 million in his next deal with the Warriors, but because he did not make the selection, he was only eligible for a 5-year $194 million deal (Davis, 2019). The snub was worth millions, but did Klay really get snubbed? Table 4.2 displays Klay's statistics for the 2018–2019 season in comparison to the other 6 guards that made the All-NBA selection that year.[1]

As shown in Table 4.2, Klay Thompson had a higher field goal percentage than 4 of the guards that made the selection, and a higher 3-point percentage than all of the other guards who made the selection except his fellow Warrior Stephen Curry. Although he did not score as many points per game as any other guard, his higher percentages show that he is much more efficient than a few of the guards on the list.

So, was Klay snubbed? Arguments can be made on why he was snubbed or why he was not, but the bigger question is: does the NBA get the All-NBA selections correct? How often does a superstar player like Klay Thompson get snubbed? To answer this question, I collected data on the All-NBA selections for 5 seasons between 2015 and 2020, which yielded 75 player observations since 15 players make All-NBA selections each year. The players were classified into first, second, or third team selections. Along with the three groups of players, a fourth group was also collected: players who had a top 15 player efficiency rating (PER) for each of the 5 seasons but had not made the All-NBA selection for each respective season.

---

[1] These statistics were collected from www.basketball-reference.com.

The PER is a statistic that considers per minute efficiency where both positive (e.g., points, assists, rebounds) and negative (e.g., turnovers, missed shots) influencers of player efficiency are considered.[2] It was important to collect data for this fourth group in order to make the comparison with the groups that had been selected for All-NBA. Table 4.3 displays this fourth group. The overall sample that included the 75 all-NBA selections along with players who did not make it but have top 15 PERs produced an overall sample of 104 players. The PER for all players in the sample were collected to be used as the dependent variable.

To test the differences between the groups (first team, second team, third team, and no team all-NBA selections) with regard to PER, a one-way ANOVA was

Table 4.3   Players who had a top 15 PER but did not make All-NBA selections

Season	Player	PER
2019–2020	Karl-Anthony Towns	26.53
2019–2020	Kyrie Irving	26.24
2019–2020	Joel Embiid	25.82
2019–2020	Hassan Whiteside	25.07
2019–2020	Zion Williamson	24.14
2019–2020	Trae Young	23.95
2018–2019	Anthony Davis	30.32
2018–2019	Karl-Anthony Towns	26.38
2018–2019	Nikola Vucevic	25.53
2018–2019	Jonas Valanciunas	24.47
2018–2019	Boban Marjanovic	24.25
2017–2018	Kyrie Irving	25.03
2017–2018	Montrezl Harrell	24.73
2017–2018	Clint Capela	24.55
2017–2018	Nikola Jokic	24.52
2017–2018	Chris Paul	24.39
2016–2017	Nikola Jokic	26.40
2016–2017	Chris Paul	26.25
2016–2017	Karl-Anthony Towns	26.00
2016–2017	DeMarcus Cousins	25.84
2016–2017	JaVale McGee	25.26
2015–2016	Boban Marjanovic	27.77
2015–2016	Hassan Whiteside	25.69
2015–2016	James Harden	25.36
2015–2016	Anthony Davis	25.10
2015–2016	Enes Kanter	24.09
2015–2016	Jonas Valanciunas	22.63
2015–2016	Karl-Anthony Towns	22.59
2015–2016	Carl Landry	22.51

[2] http://insider.espn.com/nba/hollinger/statistics.

conducted in SPSS 26. Results from the analysis showed a statistically significant difference across the groups in terms of PER: $F(3,100) = 22.11$, $p = 0.001$. These results are displayed in Table 4.4.

To test which groups actually had differences, a post hoc Tukey test was conducted in SPSS 26 and showed that the first team had significantly higher PER than the second team at the $p < 0.001$ significance level, the third team at the $p < 0.001$ level, and the no team group at the $p < 0.001$ level. The second team had significantly higher PER than the third team at the $p < 0.05$ significance level but had no significant difference in PER from the no team group. Finally, and most interestingly, the third team had significantly lower PER at the $p < 0.001$ significance level than the no team group. The results of this post hoc analysis are shown in Table 4.5.

What the results of these analyses mean are summarized with the following points:

- First team All-NBA selections have the highest PER in comparison to the other groups.
- Second team All-NBA selections have higher PER than the third team, but there is no difference between the PER scores of second team All-NBA selections and the PER scores of the 29 players who were not selected to any All-NBA teams in the 5 years between 2015 and 2020 despite having had top 15 PER scores.
- Third team All-NBA selections had lower PER scores than the 29 players who were not selected to any All-NBA teams in the 5 years between 2015 and 2020 despite having had top 15 PER scores.

When PER is considered, the All-NBA voters seemed to get first team selections correct, but second and third team selections have equal to or lower PER scores than many players who did not make the All-NBA in the 5 years between 2015 and 2020. One notable case of this for me is the case of Karl-Anthony Towns, or KAT. In the 5 years examined, KAT made the top 15 in PER every year, as was shown in Table 4.3, yet he only made the All-NBA third team in 2018. Furthermore, in the 2019–2020 season, KAT averaged 26.5 points per game on nearly 51% field goal percentage, 41% 3-point percentage, and 10.8 rebounds per game, yet he did not make All-NBA that year. Table 4.6 displays the 2019–2020 season statistics for KAT along with the 3 centers who were selected to an All-NBA team.

As shown in Table 4.6, in the 2019–2020 season, KAT had a higher field goal percentage than Anthony Davis, a higher 3-point percentage than all 3 centers who were selected All-NBA, a higher free throw percentage than Rudy Gobert, more rebounds per game than Anthony Davis and Nikola Jokic, more assists per game

Table 4.4 ANOVA results for the group differences between first, second, third team All-NBA and players who had a top 15 PER that did not make All-NBA selections

	SS	MS	df	F	p
Between groups	363.15	121.05	3	22.11	0.001
Within groups	547.42	5.47	100		
Total	910.58		103		

Table 4.5 Results from Tukey's post hoc analysis

		Mean difference	Standard error	p	95% confidence interval	
					Lower bound	Upper bound
First team	Second team	3.11	0.66	0.00	1.38	4.84
	Third team	5.34	0.66	0.00	3.61	7.07
	Not selected	2.39	0.64	0.00	0.72	4.06
Second team	First team	−3.11	0.66	0.00	−4.84	−1.38
	Third team	2.23	0.66	0.01	0.50	3.96
	Not selected	−0.72	0.64	0.68	−2.39	0.95
Third team	First team	−5.34	0.66	0.00	−7.07	−3.61
	Second team	−2.23	0.66	0.01	−3.96	−0.50
	Not selected	−2.95	0.64	0.00	−4.62	−1.28
Not selected	First team	−2.39	0.64	0.00	−4.06	−0.72
	Second team	0.72	0.64	0.68	−0.95	2.39
	Third team	2.95	0.64	0.00	1.28	4.62

Table 4.6 Statistics for KAT and the 3 centers who made an All-NBA selection in 2019–2020

Player	FG%	3P%	2P%	FT%	Rebounds per game	AST	STL	BLK	Points per game
Karl-Anthony-Towns	0.508	0.412	0.586	0.796	10.8	4.4	0.9	1.2	26.5
Anthony Davis	0.503	0.33	0.546	0.846	9.3	3.2	1.5	2.3	26.1
Nikola Jokic	0.528	0.314	0.594	0.817	9.7	7	1.2	0.6	19.9
Rudy Gobert	0.693	N/A	0.693	0.63	13.5	1.5	0.8	2	15.1

than Anthony Davis and Rudy Gobert, more steals than Rudy Gobert, more blocks than Nikola Jokic, and more points per game than all of the other 3 centers. With statistics like his during that season, why did he not get selected to an All-NBA team? Perhaps the NBA needs to make changes regarding how All-NBA selections are made.

The NBA has an All-Star selection for popular players in the league and has an All-star game to showcase such stars. All-Stars are selected every year through a voting process that includes fans, members of the sports media, and players. If the NBA already has a structure in place to honor the most popular players in the All-Star process, why does the All-NBA have to be done by votes as well? Personally, I love the All-Star process as it is meant to highlight the fun and the celebrity of players in the league. However, the All-NBA selection process should be less about the popularity of players, and more about the objective realities of who really are the best in the league in their positions. With more objective ways to select the All-NBA teams, perhaps the players who do not make the teams would not have similar or better statistics than players who make the second and third All-NBA teams each year.

4.2 Conclusion

Klay Thompson could not get a supermax contract because he did not make an All-NBA team. KAT scored 26.5 points per game on nearly 51% field goals (41% 3-pointers), and 10.8 rebounds per game, yet still did not make an All-NBA team. This is problematic. The analyses in this chapter provide evidence that statistically, players that make second All-NBA teams have similar PER as many players who did not make any All-NBA team, and players who make third All-NBA teams have lower PER than many players who did not make any All-NBA team. It would not be so bad if making All-NBA did not influence player careers and salaries, but it does, and the NBA should change this process.

Reference

Davis, S. (2019). *Klay Thompson gave an all-time eye roll and grew frustrated after learning he missed out on nearly $27 million by not making an All-NBA team.* Business Insider. Retrieved from https://www.businessinsider.com/klay-thompson-misses-supermax-contract-all-nba-2019-5

Chapter 5
Small Ball in a Big Man's Game

5.1 What Is Small Ball?

A team uses a small ball lineup when a traditional power forward is replaced with a more agile and faster small forward or shooting guard and places a traditional power forward in the center position. Players like Carmelo Anthony, Lebron James, Kevin Durant, and Eric Gordon have been used strategically at the power forward position in their respective teams to implement the small ball lineup. This style of play allows for the team to be much more versatile and increase the speed of the game. This strategy has become popular over recent years, popularized by the Golden State Warriors in their run between 2014 and 2019. However, not all teams have found success with this type of lineup. This chapter seeks to analyze when the small ball lineup works, and when it does not. Specifically, the analyses in this chapter examine the factors that allow small ballers to thrive in a big man's game. I do this by first analyzing the Golden State Warriors' success with this strategy.

5.2 Golden State's Death Lineup

In the fall of 2014, Steve Kerr commenced his endeavor as the head coach of the Golden State Warriors. He achieved immediate success on this team, lead by eventual 2-time MVP Stephen Curry, and supported by 7-foot center Andrew Bogut, superstar defender and all-star Andre Iguodala as well as eventual all-stars Draymond Green and Klay Thompson. In that season, the Warriors won 67 out of the 82 regular season games to claim the #1 seed in the league and marched into the finals to face Lebron James' Cleveland Cavaliers. After 3 games in the finals, Golden State had fallen 2-1 to the Cavaliers. Up to that point (throughout the season and the playoffs), Steve Kerr had opted to start the 7-foot-tall Andrew Bogut at center. However,

A. A. Randrianasolo, *Triple Double*, SpringerBriefs in Statistics,
https://doi.org/10.1007/978-3-030-79032-5_5

Kerr's solution to counter the Cavaliers (and Lebron's dominance), was to put Bogut on the bench and start Stephen Curry, Klay Thompson, Andre Iguodala, Harrison Barnes, and Draymond Green for Game 4. In this lineup, Draymond Green was moved to the center position, which is remarkable not only because he would have to face the Cavaliers' 7′1″ center, Timofey Mozgov, but also because he was shorter than Harrison Barnes and the same height as Klay Thompson and Andre Iguodala on his own team. The Cavaliers' starting lineup was not only led by the 6′9″ greatest player on earth known as Lebron James but were remarkably bigger than the Warriors, as shown in Table 5.1.

Steve Kerr went with this starting lineup while down 2-1 in the finals and saw success. The Warriors beat the Cavaliers 103-82. What is especially impressive by this turnout was not just that the Warriors had beaten a much larger team by 21 points, it was that they held this team to only 82 points. The Cavaliers had scored 100, 95, and 96 points in the previous 3 games in the series. They were able to do this by speeding up the pace of the game, spreading the floor, and finding open shots. Furthermore, their defensive versatility greatly contributed to this win as multiple players in the starting lineup were able to switch on and off of guarding multiple positions.

Kerr kept this lineup for games 5 and 6, and the Warriors would win the championship in 6 games. The turning point of the series was definitely the decision to play small ball in Game 4. This decision is known to have birthed the Warriors' "Death Lineup" that ultimately lead to 5 finals appearances and 3 championships in the subsequent years.

In the summer of 2016, after losing to the Cavaliers in the finals, four Golden State Warriors players (i.e., Stephen Curry, Klay Thompson, Andre Iguodala, and Draymond Green) traveled to the Hamptons in New York to recruit Kevin Durant to come join their team. Durant agreed, and thus the "Hamptons 5" lineup was created. Over the subsequent years, with addition of the Durantula, the Warriors' small ball basketball system would win 67, 58, and 58 games in the regular seasons, appear in the finals 3 times and win 2 championships. Clearly, the Warriors had figured out how to win with the small ball system.

Not all teams have figured out how to win with the small ball system however most notably the Houston Rockets. In February 2020, the Rockets traded away their 6′10″ center, Clint Capela and acquired 6′7″ Robert Covington. In that season, the

Table 5.1 Height and position of the Golden State Warriors and Cleveland Cavaliers' starting lineups in the 2014–2015 NBA finals

	Golden State Warriors		Cleveland Cavaliers	
Position	Player	Height	Player	Height
Guard	Stephen Curry	6′3″	Matthew Dellavedova	6′3″
Guard	Klay Thompson	6′6″	Iman Shumpert	6′5″
Forward	Harrison Barnes	6′8″	Lebron James	6′9″
Forward	Andre Iguodala	6′6″	Tristan Thompson	6′9″
Center	Draymond Green	6′6″	Timofey Mozgov	7′1″

Table 5.2 2019–2020
Houston Rockets small
ball lineup

Player	Height
Russell Westbrook	6′3″
James Harden	6′5″
Eric Gordon	6′3″
Robert Covington	6′7″
P.J. Tucker	6′5″
Danuel House	6′6″
Jeff Green	6′8″
Ben McLemore	6′3″

mighty Warriors and their Hamptons 5/Death Lineup, who had blocked the Rockets' path to the championship in 3 out of the previous 4 seasons in the playoffs, were no longer a threat as Kevin Durant had left the team to join the Brooklyn Nets, and both Stephen Curry and Klay Thompson were recovering from injuries. With the Warriors out of the picture, the Rockets fully embraced their small ball lineup, led by two former MVPs: James Harden and Russell Westbrook. Table 5.2 displays the Rockets' small ball lineup in that season. After the Clint Capela trade, the Rockets' starting lineup was composed of a configuration of 5 out of 8 players shown in Table 5.2.

With the small ball lineup ready to go, the Warriors out of the picture, and two former MVPs hungry for a championship, the Rockets went full steam ahead into the playoffs. Their run was cut short however by the Lakers in the second round of the playoffs. The Lakers took care of the Houston Rockets in just five games. It seemed that the Rockets had still not figured out how to be effective with the small ball lineup.

Although the Warriors have found tremendous success with the small ball strategy, not every team has done the same. This means that there are certain conditions under which the small ball method works. This chapter investigates these conditions.

5.3   When Does Small Ball Work?

To answer this question, I collected data on "small ball" teams between 2014 and 2020. To determine the small ball teams, a review of popular sports media was conducted, and the small ball teams named from this review (e.g., Buckley, 2020; Davis, 2017; Martin, 2015; Trebisovsky, 2018; Warond, 2017; Whitney, 2019) were included in the analysis. This yielded 19 teams, as shown below in Table 5.3.

General managers and coaches opt to implement the small ball strategy for several reasons. First, small ball allows for a more up-tempo game. The advantage in this is that up-tempo games allow for smaller players, quicker players to thrive against larger and slower players. This creates more open lanes in transition. Second, it can allow for an offense to spread the floor and create more open shots, as smaller

Table 5.3 Small Ball teams
from 2014 to 2020

Team	Season
Boston Celtics	2019–2020
Cleveland Cavaliers	2017–2018
Dallas Mavericks	2014–2015
Golden State Warriors	2014–2015
Golden State Warriors	2015–2016
Golden State Warriors	2016–2017
Golden State Warriors	2017–2018
Golden State Warriors	2018–2019
Houston Rockets	2019–2020
Houston Rockets	2017–2018
Indiana Pacers	2014–2015
Los Angeles Clippers	2019–2020
Miami Heat	2019–2020
Miami Heat	2014–2015
OKC Thunder	2019–2020
OKC Thunder	2017–2018
Philadelphia 76ers	2017–2018
San Antonio Spurs	2017–2018
Charlotte Hornets	2019–2020

players can pop in and out of the paint more efficiently than larger forwards and centers. Finally, the small ball method can allow for more defensive versatility as players can switch from guarding forwards to guards, which is needed to defend the pick and roll from many offenses. As previously discussed however small ball does not always work. So, when does it work?

Well, in the current NBA, it works when the ball is moved around more to create assists for 3 pointers. Again, small ball systems are built to spread the floor, move the ball around, and create open outside shots. To provide evidence for this, data was collected on the 19 teams in Table 5.3. For each team, 3-point field goal percentage per game, points per game, winning percentage, and assists per game were collected.[1] It is important to note that what I am proposing here is that small ball is effective when the team can spread the floor for effective assisted 3 pointers. In other words, small ball is effective when the ball is moved around and 3-point shooters are assisted. This proposition suggests that in the small ball system, 3-point percentage mediates the relationship between assists and winning percentage, as shown in Fig. 5.1.

To test this mediation, the PROCESS macro in SPSS 26 was utilized with 5000 bootstraps. Since this is a simple mediation, model 4 was used to run the analysis. The results showed that 3-point percentage mediates the relationship between assists and winning percentage, as shown in Table 5.4.

---

[1] All statistics were collected from www.basketball-reference.com.

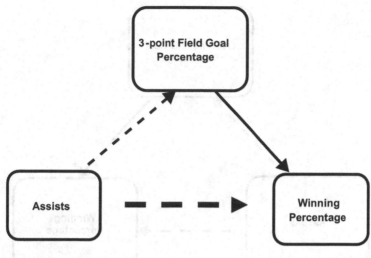

Fig. 5.1 Proposed mediation 1

Table 5.4 PROCESS macro results for the mediation of 3-point percentage in the relationship between assists and winning percentage

	Effect	Standard error	t-value	p-value	CI (95%) Lower	Upper
Direct and total effects						
3-point percentage --> Winning Percentage	6.34	1.96	3.23	0.005	2.181	10.494
Assists --> Winning Percentage	−0.009	0.0121	−0.735	0.473	−0.034	0.17
Indirect effects						
Assists --> 3-point percentage--> Winning percentage	0.03	0.01			0.013	0.07
R Square: 0.62, *p* < 0.001						

What Table 5.4 shows is that:

1. For small ball teams, assists do not directly influence winning percentage
2. For small ball teams, 3-point percentage influences winning percentage
3. For small ball teams, the mediation of 3-point percentage in the relationship between assists and winning percentage is significant

What can be stated from these results is that within the small ball teams, assisted 3-pointers are key to winning. However, could it not just be that passing the ball leads to better shooting spots, which may not necessarily be 3-pointers? To clear this up, a mediation analysis was conducted to see if field goal percentage (rather than just 3-point percentage) mediates the relationship between assists and winning percentage. The mediation model is shown in Fig. 5.2.

To test this mediation, the PROCESS macro in SPSS was again utilized with 5000 bootstraps. The results showed that overall field goal percentage does not

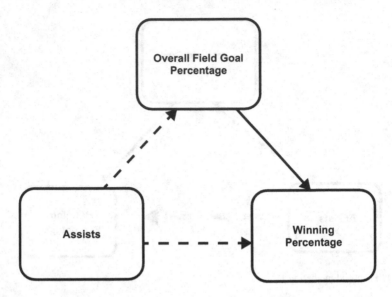

Fig. 5.2 Proposed mediation 2

Table 5.5 PROCESS macro results for the mediation of field goal percentage in the relationship between assists and winning percentage

	Effect	Standard error	t-value	p-value	CI (95%) Lower	Upper
Direct and total effects						
Field goal percentage --> Winning Percentage	5.52	2.13	2.59	0.02	.996	10.041
Assists --> Winning Percentage	−0.0003	0.011	−0.024	0.98	−0.025	0.0245
Indirect effects						
Assists --> Field goal percentage--> Winning percentage	0.02	0.014			−0.0008	0.0464
R Square: 0.74, *p* < 0.01						

mediate the relationship between assists and winning percentage, as shown in Table 5.5.

What Table 5.5 shows is that:

1. For small ball teams, assists do not directly influence winning percentage
2. For small ball teams, field goal percentage positively influences winning percentage
3. For small ball teams, the mediation of field goal percentage in the relationship between assists and winning percentage is not significant

These results show that although field goal percentage does affect winning for small ball teams, there is no significance in the mediation of field goal percentage in the relationship between assists and winning percentage. What this tells us is that the small ball strategy is not effective with assisted field goals in general, but specifically only effective with assisted 3-point field goals. To provide further justification for this stance, one more analysis was conducted. In this final analysis, the mediation of 2-point field goal percentage in the relationship between assists and winning percentage is tested, as shown in Fig. 5.3.

Again, the PROCESS macro in SPSS was utilized with 5000 bootstraps. The results showed that 2-point field goal percentage does not mediate the relationship between assists and winning percentage, as shown in Table 5.6.

What Table 5.5 shows is that:

1. For small ball teams, assists do not directly influence winning percentage
2. For small ball teams, 2-point field goal percentage positively influences winning percentage
3. For small ball teams, the mediation of 2-point field goal percentage in the relationship between assists and winning percentage is not significant

These results show that although field goal percentage does affect winning for small ball teams, there is no significance in the mediation of field goal percentage in the relationship between assists and winning percentage. What this tells us is that the small ball strategy is not effective with assisted 2-point field goals, but specifically only effective with assisted 3-point field goals. The following paragraphs discuss the strategic implications of the findings of the analyses conducted.

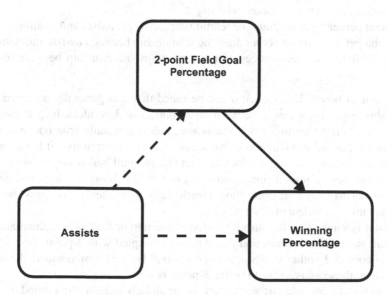

Fig. 5.3 Proposed mediation 3

Table 5.6 PROCESS macro results for the mediation of 2-point field goal percentage in the relationship between assists and winning percentage

	Effect	Standard error	t-value	p-value	CI (95%) Lower	Upper
Direct and total effects						
2 Point percentage --> Winning Percentage	2.302	1.01	2.29	0.04	0.1701	4.433
Assists --> Winning Percentage	0.0157	0.008	1.98	0.07	−0.0011	0.0325
Indirect effects						
Assists --> 2 Point percentage--> Winning percentage	0.001	0.007			−0.0028	0.0220
R Square: 0.74, $p < 0.01$						

5.4   What Does This Mean?

Small ball can be extremely effective, as we all saw with the Golden State Warrior's run. However, this method of play can also be ineffective for some teams, as seen with the Houston Rockets during the Warriors' run. The analyses conducted in this chapter provide some guidance as to when this style of play can be effective, and the results show some specific findings with respect to small ball teams:

1. 3-point percentages positively influence winning
2. 2-point percentages positively influence winning
3. Overall field goal percentages positively influence winning
4. Assists alone do not influence winning
5. 3-point percentages mediate the relationship between assists and winning
6. 2-point percentages do not mediate the relationship between assists and winning
7. Overall field goal percentages do not mediate the relationship between assists and winning

Looking at points 1, 2, and 3; it can be stated that it is generally accepted that better shooting percentages, whether it be 2 pointers or 3 pointers, help teams win games. Yes, better shooting helps teams win, this is generally true not only with small ball teams but with all teams. Although this is pretty intuitive, it is important to establish this base in order to discuss what makes small ball strategies work. The small ball strategy allows teams to move at a faster pace through transitions, spread the floor out to help find open shots (particularly 3 pointers), and pass the ball around to more mobile and agile players.

Point 4 is interesting because it reinforces the notion that assists alone are not sufficient in small ball teams, and that it must be coupled with 3-point shooting, as stated in point 5. In other words, it is not enough to have a lot of assists in the small ball system, those assists should be for 3-pointers.

Points 6 and 7 provide further support for the already stated notion found in point 4. By finding that field goal percentage and 2-point field goals do not mediate the relationship between assists and winning percentage, it can further be stated that the small ball style of play needs assisters who find 3-point shooters, rather than assisters who assist 2-pointers (e.g., lobs and alley-oops).

So, let us go back to the examples discussed earlier in this chapter. As previously mentioned, the Golden State Warriors switched to a small ball lineup in Game 4 of the 2014–2015 NBA finals after being down 2-1 to the Cleveland Cavaliers and ended up winning the championship. The 2019–2020 Houston Rockets also used the small ball lineup but could not get past the Lakers in the 2020 western conference semifinals. Table 5.7 displays the assists, 3 point percentages, and points scored in these games.

As shown in Table 5.7, in Game 4 of the 2014–2015 finals, Golden State was able to have 24 assists and shoot 40% from the 3-point line (with the splash brothers), while the Cavaliers had only had 16 assists and shoot 14.8% from beyond the arc. Here, the Warriors could be effective with the small ball lineup because they had more assisted 3-pointers than the Cavs.

In Game 1 of the 2019–2020 western conference semifinals, the Rockets shot 35.9% from 3, while the Lakers only shot 28.8%, and also the Rockets had 19 assists to the Lakers' 18, allowing the Rockets to win the game. In Games 2 and 5 however the Rockets had less assists and a lower 3-point percentage than the Lakers, and thus they lost those games. In Games 3 and 4, the Rockets had higher 3-point percentages than the Lakers, but less assists, and thus lost both games. This further confirms that higher 3-point percentages alone are not sufficient for small ball, they must be accompanied with assists.

Table 5.7 3-point percentage and assists for Game 4 of the 2014–2015 NBA finals, and the 2019–2020 Western Conference semifinals

	Assists	3-point percentage	Points
Game 4 2014–2015 NBA finals			
Golden State Warriors	24	40%	**103**
Cleveland Cavaliers	16	14.8%	82
Game 1 2019–2020 Western Conference semifinals			
Houston Rockets	19	35.9%	**112**
Los Angeles Lakers	18	28.8%	97
Game 2 2019–2020 Western Conference semifinals			
Houston Rockets	22	41.5%	109
Los Angeles Lakers	30	44.4%	**117**
Game 3 2019–2020 Western Conference semifinals			
Houston Rockets	23	40%	102
Los Angeles Lakers	25	33.3%	**112**
Game 4 2019–2020 Western Conference semifinals			
Houston Rockets	24	42.4%	100
Los Angeles Lakers	30	30%	**110**
Game 5 2019–2020 Western Conference semifinals			
Houston Rockets	18	26.5%	96
Los Angeles Lakers	25	51.4%	**119**

From the findings of the analyses in this chapter, I can make two recommendations to coaches or general managers seeking to implement the small ball strategy:

- **Recommendation 1:** If a team wants to implement the small ball method, they need to make sure that they have good passers on the team that can find open 3-point shooters.
- **Recommendation 2**: If a team wants to implement the small ball method, they need to make sure that they have good 3-point shooters.

5.5 Conclusion

It has always been amazing to me how the success of one team in the NBA can have such a great influence on how other teams construct their lineups. It seemed as if as soon as Golden State's Death Lineup small ball method clicked, lots of other teams tried to mimic this strategy and started to construct smaller lineups. The 2017–2018 Oklahoma City Thunder tried to copy this strategy with their Russell Westbrook/Paul George/Carmelo Anthony big 3, yet only won 48 games that season and got bounced out in the first round of the playoffs. Similarly, the 2019–2020 tried to implement a smaller lineup lead by one of the most prolific scorers in the game, James Harden, and one of the most explosive players in the game, Russell Westbrook, but ultimately fell 4-1 to the Lakers in the western conference semifinals. What these teams are missing is one thing: Golden State was able to find success with their Death Lineup in 2015 because of the splash brothers. Steph Curry is not only an effective passer but is regarded by many as the greatest 3-point shooter in history. Klay Thompson is a career 42% 3-point shooter. When they brought Kevin Durant into the mix, they added a career 38% 3-point shooter. If you want to use small ball, make sure your players can pass the ball and make it rain from beyond the arc.

References

Buckley, Z. (2020). *Ranking the NBA's best small ball lineups this season*. Bleacher report. Retrieved from https://bleacherreport.com/articles/2884477-ranking-the-nbas-best-small-ball-lineups-this-season

Davis, S. (2017). *The rockets pushed the NBA's small-ball era to a new extreme in their first game of the season*. Insider. Retrieved from https://www.insider.com/rockets-small-ball-win-vs-warriors-2017-10

Martin, J. (2015). *Projecting the NBA's best small-ball teams*. Bleacher Report. Retrieved from https://bleacherreport.com/articles/2568498-projecting-the-nbas-best-small-ball-teams

Trebisovsky, J. (2018). *Rockets GM, Daryl Morey, combines love for basketball and theater with new musical 'SMALL BALL'*. USA Today. Retrieved from https://rocketswire.usatoday.com/2018/03/28/rockets-gm-daryl-morey-combines-love-for-basketball-and-theater-with-new-musical-small-ball/

Warond, A. (2017). *Ranking the top 6 best small ball lineups in the NBA*. Fadeaway World. Retrieved from https://fadeawayworld.com/2017/11/15/ranking-the-top-6-best-small-ball-lineups-in-the-nba/

Whitney, C. (2019). *Lineup series: Small ball*. SB Nation—All the Hive. Retrieved from https://www.atthehive.com/2019/10/20/20907016/lineup-series-small-ball

Chapter 6
Is the Clutch Gene Real?

6.1 Who Is Clutch?

On June 14, 1998, 11 days before my 11th birthday, Michael Jordan was set to play game 6 of the 1998 NBA finals against the Utah Jazz. I watched this game with my dad (and my little brother who had fallen asleep on the couch), and I remember even before the game, the mystique and aura around his Royal Airness was at an all-time high. As I watched the game wind down to within the final minute, I realized that the Bulls were down 85–86, and they needed a bucket to win. Well, who better to have on your team than #23 when you need a bucket? The stars aligned that night because right around 20 s left in the game, Michael Jordan steals the ball from Karl Malone, dribbles down the court, pulls off an almighty crossover on Bryon Russell, and sinks the shot to put the dagger into the Jazz's hopes of holding the trophy. I jumped up off of the couch and screamed in joy, meanwhile, my dad told me to keep it down, so I do not wake my brother from his slumber. This was the last game Jordan would play as a Bull, and this shot cemented his legacy as the most clutch player the game has ever seen.

Fast forward to June 19, 2016, 18 years later and 6 days before my 29th birthday, Lebron James along with Kyrie Irving and the rest of the Cleveland Cavaliers were set to face Stephen Curry and the 73-win Warriors in game 7 of the NBA finals. I was living in Cleveland at the time as a professor at John Carroll University, but I traveled to St. Louis to spend time with family and watch the NBA finals with my little brother (a different brother from the 1998 finals, but still sleeps a lot). I was a huge fan of the Warriors as I followed Steph's career since he was at Davidson College, and I really wanted him to beat the Cavs again. I was probably the only person in Cleveland who was rooting for the Warriors (my friends constantly reminded me of how much of a traitor I was). As the game wound down to the final minute, the score was tied 89 to 89, and Cleveland had the ball. The ball was in Kyrie Irving's hands and the Cavs had run a pick and roll to get Steph rather than

A. A. Randrianasolo, *Triple Double*, SpringerBriefs in Statistics, https://doi.org/10.1007/978-3-030-79032-5_6

Klay Thompson who was an excellent defender to guard Kyrie. Kyrie dribbles for a few moments and sinks a 3-pointer in Steph Curry's face to put the Cavs 3 points ahead. Many people remember this shot, but they fail to remember that after this shot, the game was not over. There were still about 45 s left on the clock and the Warriors had two of the greatest shooters to ever live on their team. In the possession after Kyrie's big 3-pointer over Steph, Steph had a chance to equalize the score yet missed his 3-pointer. The Cavs would go on to finish the game and bring the title back to Cleveland. So, Kyrie hits a big time clutch 3-pointer, while Steph had missed his shot, and did not seem as clutch.

Fast forward 3 more years, and now it is Steph Curry and the Warriors in the finals once again, but this time against Kawhi Leonard and the Toronto Raptors. It was June 13, 2019, and it was game 6 in the series. The Warriors had their backs against the wall, as they faced elimination with a loss. To make matters worse, during the game, two of the Warriors' superstar players had gone down with catastrophic injuries: Kevin Durant had torn his Achilles tendon and Klay Thompson suffered an ACL injury. It seemed as if the weight of the Warriors' championship hopes had fallen onto the shoulders of the greatest shooter in the world, Stephen Curry. The game had come down to the wire. With 9.6 s left in the game, the Warriors were down 111–110, but they had possession of the ball. This was Steph Curry's chance to cement his legacy as the greatest shooter ever. The Warriors inbounded the ball and got it to Steph, but unfortunately for the dubs, Steph missed the shot, and the Raptors would go on to win the championship. Steph Curry had lost twice in the finals, and in both losses, he had missed a crucial 3-pointer in the waning seconds of the last game in the series. Does this mean that Steph is not clutch, or that he does not have the "clutch gene"?

On ESPN's debate show, First Take, Max Kellerman famously stated that if the fate of the universe is on the line, he would rather have Andre Iguodala take the last shot in a game than Steph Curry (the subsequent memes that came from Max's declaration were hilarious, by the way). We are talking about Steph Curry, the 2-time MVP, the man who shot over 41% from beyond the arc in his first 10 seasons in the NBA, the man who led the leagues in 3 pointers for 5 seasons, and the man who has the NBA record for most 3-pointers in one season with 402. No disrespect to Andre Iguodala's greatness, but if I need a shot to go in at any point during any game, I probably will rather go with Steph. But whether you would take Steph 3-pointer, or place your trust in Iguodala's hands with the fate of the universe on the line, the question that remains is: does the clutch gene exist? In other words, do Kyrie Irving and Michael Jordan possess some sort of clutch ability or trait that Steph Curry does not? This chapter focuses on answering this question with statistical analyses.

6.2 Analyzing Clutch Statistics

In order to test the clutch gene, the term "clutch" must first be defined. In sports, clutch refers to a player's ability to perform under circumstances where there is heightened pressure to perform well, such as situations where the player's team needs a bucket in the waning seconds of a playoff game. Therefore, I define clutch here as: *the ability of a player to perform well under high pressure situations.* So, is this trait real?

To investigate whether or not this trait is real, I collected data on clutch player shooting from the 2019–2020 playoffs. Playoff games have much more at stake than regular season games, and therefore is more appropriate to analyze with respect to the clutch gene than regular season games. Inpredictable.com provides statistics on player shots in the playoffs in different categories.[1] This site provides the top 50 clutch shooters in the playoffs for each season, and the following statistics were collected from this site for the analysis in this chapter:

- Field goal percentage for normal shots: this is defined on the site as shots where the probability impact is normal and encompasses the majority of shots.
- Field goal percentage for clutch 1 shots: shots that have an elevated impact on the shot-taker's team winning the game.
- Field goal percentage for clutch 2 shots: shots that are crucial to the game's outcome. These include buzzer beater (or potential buzzer beater) shots at the end of a game.

From the 50 players on the list, 18 were deleted from the dataset because they had not taken any clutch 2 shots. Specifically, these players shot plenty of clutch 1 shots, yet no clutch 2 shots, and in order to be fair and consistent, they were deleted in order to examine both clutch 1 and clutch 2 shots. This left a total of 32 observations. Table 6.1 displays the clutch statistics for the players in the dataset.

To analyze the data, the z scores for the three variables were first calculated in SPSS 26. It is important to use the standardized z scores in this analysis to be able to compare the variables appropriately. Then, 2 paired sample t tests were employed to test the player's shooting percentage under different conditions. The first paired sample t test was used to test if there is a significant difference between player shooting percentages in normal shots and in clutch 1 shots. The results found no significant difference in shooting percentage for normal shots ($M = 0.53$, $SD = 0.06$) and clutch 1 shots ($M = 0.52$, $SD = 0.17$).

The second paired sample t test was used to test if there is a significant difference between player shooting percentages in normal shots and in clutch 2 shots. The results found no significant difference in shooting percentage for normal shots ($M = 0.53$, $SD = 0.06$) and clutch 1 shots ($M = 0.54$, $SD = 0.45$). The results of the 2 *t* tests are displayed in Table 6.2.

---

[1] http://stats.inpredictable.com/nba/ssnPlayerSplit.php.

Table 6.1 Clutch statistics for players in the 2019–2020 playoffs

Player	Normal shot %	Clutch 1%	Clutch 2%
Jamal Murray	0.586	0.521	0.333
LeBron James	0.631	0.571	0.5
Jimmy Butler	0.516	0.47	0.75
Tyler Herro	0.488	0.742	1.5
Kemba Walker	0.513	0.609	0.4
Jayson Tatum	0.51	0.531	0
Goran Dragic	0.534	0.5	0
Nikola Jokic	0.573	0.741	0.75
Jaylen Brown	0.561	0.577	0.5
Bam Adebayo	0.56	0.654	1
Dennis Schroder	0.487	0.4	0.667
Fred VanVleet	0.523	0.312	0.167
Kawhi Leonard	0.547	0.619	0
Donovan Mitchell	0.622	0.643	0.667
Marcus Smart	0.471	0.625	0.75
Chris Paul	0.541	0.6	0.4
Khris Middleton	0.5	0.194	0.333
Pascal Siakam	0.444	0.367	0.333
Daniel Theis	0.5	0.667	1
Anthony Davis	0.599	0.567	1.167
Kentavious Caldwell-Pope	0.551	0.567	0
Russell Westbrook	0.47	0.267	0.333
James Harden	0.558	0.433	0
Paul George	0.491	0.538	0.5
Shai Gilgeous-Alexander	0.559	0.154	1.5
Lou Williams	0.452	0.375	0.5
Norman Powell	0.477	1	0.333
Danny Green	0.419	0.5	0
Mike Conley	0.641	0.444	0
Rudy Gobert	0.635	0.667	1
Brook Lopez	0.632	0.5	0.5
OG Anunoby	0.523	0.375	1.5

Table 6.2 Results for paired sample t tests for 2019–2020 sample

	M	SD	t	p
Analysis 1			0.394	0.696
Normal shots	0.53	0.06		
Clutch 1 shots	0.52	0.17		
Analysis 2			−0.105	0.917
Normal shots	0.53	0.06		
Clutch 2 shots	0.54	0.45		

Table 6.3 Clutch statistics for players in the 2018–2019 playoffs

Player	Normal shot %	Clutch 1%	Clutch 2%
Kawhi Leonard	0.547	0.51	0.556
Stephen Curry	0.561	0.5	0.167
CJ McCollum	0.521	0.465	0.4
Klay Thompson	0.572	0.371	0.75
Jamal Murray	0.468	0.588	0
Damian Lillard	0.524	0.561	0.321
James Harden	0.519	0.5	0
Nikola Jokic	0.539	0.404	0
Eric Gordon	0.582	0.58	0
Pascal Siakam	0.521	0.396	0.6
Kyle Lowry	0.523	0.674	0
Kevin Durant	0.606	0.545	0
Will Barton	0.337	0.675	0.5
Brook Lopez	0.539	0.794	0.5
Donovan Mitchell	0.387	0.412	0
Draymond Green	0.517	0.656	0.5
Andre Iguodala	0.563	0.533	1.5
DeMar DeRozan	0.462	0.533	0
Marc Gasol	0.486	0.679	0
Gary Harris	0.542	0.571	1
Joel Embiid	0.436	0.625	0.2
Lou Williams	0.481	0.542	1
Kevon Looney	0.685	0.636	1
Tobias Harris	0.516	0.25	1
Jimmy Butler	0.504	0.55	1
P.J. Tucker	0.64	0.2	1
Enes Kanter	0.542	0.222	0
Danny Green	0.53	0.188	0
Evan Turner	0.329	0.375	0
Seth Curry	0.5	0.562	1.5
Khris Middleton	0.503	0.5	0.333
JJ Redick	0.561	0.571	1.5
Ben Simmons	0.628	0.429	0
Meyers Leonard	0.667	0.286	1
Rodney Hood	0.544	0.429	1.167
Ricky Rubio	0.405	0.786	0
Clint Capela	0.557	0.571	0

Table 6.4 Results for paired sample *t* tests for 2018–2019 sample

	M	SD	t	p
Analysis 1			0.394	0.696
Normal shots	0.52	0.07		
Clutch 1 shots	0.50	0.15		
Analysis 2			−0.105	0.917
Normal shots	0.52	0.07		
Clutch 2 shots	0.47	0.51		

The 2019–2020 playoff games were played in the "NBA bubble" without fans and in different circumstances, and therefore may have distorted the results. So, the same data was collected for the 2018–2029 playoffs. After deleting players who did not take any clutch 2 shots, as was done with the previous sample, there remained 37 player observations. Table 6.3 displays these players and statistics.

To analyze the data, the z scores for the three variables were again calculated in SPSS 26. Then, 2 paired sample t tests were employed to test the player's shooting percentage under different conditions. The first paired sample t test was used to test if there is a significant difference between player shooting percentages in normal shots and in clutch 1 shots. The results found no significant difference in shooting percentage for normal shots ($M = 0.52$, SD = 0.08) and clutch 1 shots ($M = 0.50$, SD = 0.15).

The second paired sample t test was used to test if there is a significant difference between player shooting percentages in normal shots and in clutch 2 shots. The results found no significant difference in shooting percentage for normal shots ($M = 0.52$, SD = 0.08) and clutch 1 shots ($M = 0.47$, SD = 0.51). The results of the 2 *t* tests are displayed in Table 6.4.

The results of the analyses in this chapter show that:

- There is no significant difference between player normal shooting percentage and clutch 1 shooting percentages.
- There is no significant difference between player normal shooting percentage and clutch 2 shooting percentages.

The bottom line here is that statistically, there seems to be no evidence of a clutch gene. The likelihood of a player making a "clutch" shot is just dependent on how good of a shooter the player is. We all witnessed Stephen Curry miss 2 "clutch" shots in the NBA finals within recent years, and this may not mean that he does not have some "clutch gene," but maybe simply that he missed shots from a range where even the best shooters only make them 40% of the time.

6.3   Conclusion

Sports analyst and TV personality Skip Bayless is famous for criticisms on Lebron James' performance on the court throughout Lebron's career. His most infamous declaration relating to Lebron's performance is his declaration that in "clutch" moments, the chosen one becomes the frozen one. For the better part of the last decade, Skip, along with many Lebron critics have held that one of the things that separates Lebron from guys like Kobe Bryant and Michael Jordan is that Lebron just does not have the clutch gene. Perhaps Lebron may have had some failures in his career, but the analyses in this chapter provide evidence that not having the clutch gene is not a contributing factor to those failures. Similarly, Stephen Curry is regarded as one of the best shooters, if not the best shooter, the game has ever seen, yet his clutch perception went down a few points after missing the final shot in the NBA finals. The results of the analyses in this chapter imply that with the fate of the universe on the line, Steph is probably still the best bet to take any shot.

Chapter 7
Offense Wins Games but Does Defense Win Championships?

7.1 Defense Wins Championships!

While I was in my Ph.D. program at Saint Louis University, I decided to volunteer to be a basketball coach for my little brother's team at St. Margaret of Scotland, the grade school I had also attended. We held practice on Monday and Wednesday nights, and I was what some would call, a "defensive-minded" coach. Every Wednesday and Monday night, I drilled it into my team's understanding that we had to be excellent defenders. I had this idea that a great defense can help the team overcome any obstacle we would face in games. In other words, fundamental defense is the most important part of succeeding on the court. There were times when I did not even allow the team to touch a basketball, as we practiced defensive formations.

This notion of the importance of defense was instilled in me through my upbringing while playing all kinds of sports in little league and in high school. In fact, one of my favorite coaches used to always tell me, "offense wins games, but defense wins championships." I am sure that he was not the only coach to say this in the world, but this stuck with me for a long time, and I approached sports in this manner. However, it must be asked: what does this mean? Can a great defense truly overcome a great offense to win championships? In this chapter, I investigate this question with regard to the NBA.

A. A. Randrianasolo, *Triple Double*, SpringerBriefs in Statistics, https://doi.org/10.1007/978-3-030-79032-5_7

7.2   Analyzing Offense and Defense

To answer this question, I collected statistics for each team's performance for each game that was played in the NBA finals over 11 years.[1] The 11-year span included the NBA Finals games between 2010 and 2020. This included 22 teams (2 per year in the finals), and 64 games. Each game had 2 teams that played so that resulted in a total of 128 observations. Offensive statistics reflecting field goal percentage, 3-point shooting percentage, assists, and offensive rebounds were collected. Similarly, defensive statistics reflecting defensive rebounds, steals, and blocks were collected to reflect defensive performance in each game. The dependent variable in this analysis was whether each team won the finals series and became champions. The dependent variable was therefore categorical (winners and losers). Teams that won the championship in their respective season were coded 1, while teams that lost the series were coded 0. Table 7.1 displays the teams and in the sample.

Since the dependent variable is binary (won the finals series vs. lost the finals series), a logistic regression analysis in SPSS 26 was conducted to test the influence of offensive factors (field goal percentage, 3-point percentage, assists, offensive rebounds) and defensive factors (steals, blocks, defensive rebounds) on winning championships in the NBA finals. The logistic regression allowed for testing the influence of continuous independent variables on the categorical dependent variable. The results of the binary logistic regression found that only field goal percentage, 3-point percentage, and defensive rebounds significantly influenced winning games in the NBA finals (Cox and Snell R Square: 0.40). Table 7.2 displays the results of the analysis.

These results show that:

1. There are two offensive variables that contribute to winning: field goal percentage and 3-point percentage. Overall shooting percentage was a much stronger predictor of winning finals games than 3-point shooting percentage.
2. Although there was only one defensive variable (defensive rebounds) that significantly affected winning, this variable was a stronger predictor of winning than either of the two offensive variables.

There are several interesting findings from these results that should be noted. First, it is interesting to note that offensive rebounding is not as important to winning as defensive rebounds. This means that defensive efforts on the court in terms of rebounding is a key contributor to winning games in the finals. Perhaps this is because a defensive rebound allows the team to regain possession of the ball, and commence an opportunity to score, while offensive rebounds keep the same possession of the ball.

Second, it is also interesting to see that overall shooting percentage is more impactful than 3-point shooting in the finals. This is interesting because in the

---

[1] All statistics for the sample in this chapter were collected from https://www.espn.com/nba/stats.

Table 7.1 Teams that played in the NBA finals between 2010 and 2020

Season	Teams	Number of games played in finals	Winner
2019–2020	Miami Heat vs. Los Angeles Lakers	6	Lakers
2018–2019	Toronto Raptors vs. Golden State Warriors	6	Raptors
2017–2018	Cleveland Cavaliers vs. Golden State Warriors	4	Warriors
2016–2017	Cleveland Cavaliers vs. Golden State Warriors	5	Warriors
2015–2016	Cleveland Cavaliers vs. Golden State Warriors	7	Cavaliers
2014–2015	Cleveland Cavaliers vs. Golden State Warriors	6	Warriors
2013–2014	Miami Heat vs. San Antonio Spurs	5	Spurs
2012–2013	Miami Heat vs. San Antonio Spurs	7	Heat
2011–2012	Miami Heat vs. Oklahoma City Thunder	5	Heat
2010–2011	Dallas Mavericks vs. Miami Heat	6	Mavericks
2009–2010	Los Angeles Lakers vs. Boston Celtics	7	Lakers

Table 7.2 Results of logistic regression ($n = 128$)

Variable	B	Standard error	p
Offensive variables			
FG %	0.24	0.06	<0.001
3 point %	0.07	0.03	0.023
Assists	0.01	0.05	0.906
Offensive rebounds	0.11	0.08	0.151
Defensive variables			
Steals	0.15	0.09	0.093
Blocks	−0.05	0.08	0.535
Defensive rebounds	0.28	0.07	<0.001

current era of the NBA, teams are increasingly relying on 3-point shooting to win. Perhaps it may be better to ensure higher percentage shots over 3-pointers.

Finally, it is interesting that in the finals, steals, blocks, and assists do not significantly predict winning finals games. The findings of the analyses in this chapter find that both offense and defense win championships. Perhaps instead of saying "offense wins game, and defensive wins championships," coaches should just state that effective play on both ends of the floor wins championships.

7.3 Conclusion

Mike D'Antoni has coached several notable teams in the NBA. He was the head coach of the Phoenix Suns between 2003 and 2008. D'Antoni stepped in as the Suns' head coach in the middle of the 2002–2003 season and although that year, the team had a losing record, he was able to gain 62 wins the following season. He would go on to coach the Suns for the next 3 seasons and win no less than 65% of the games each season. During this tenure, he elevated point guard Steve Nash's game to help him win two MVP trophies (pretty impressive with guys like Kobe and Lebron in the league). However, D'Antoni's Suns would never make the NBA finals, even with an MVP. After his stint on the Suns, D'Antoni would be the head coach of the New York Knicks. Again, he would elevate the Knicks from a team that won 28% of their games in the season before he got there, to an above 0.500 playoff team, though he would never get past the first round of the playoffs with this team. After he left the Knick, D'Antoni would have an unsuccessful stint as the Lakers head coach for 2 seasons and eventually land as the Houston Rockets head coach in 2016. As the Rocket's head coach, he did not have a season with less than a 60% winning percentage and even helped superstar guard James Harden win an MVP trophy in 2018. However, even with an MVP (and 2 former MVPs in the 2019–2020 season), D'Antoni was not able to take the team to the NBA Finals. So why is D'Antoni so successful in leading his teams to the playoffs but is never able to get any gold? Perhaps because D'Antoni is known to be an offensive-minded coach that does not quite emphasize the importance of defense on his teams. Offense may be important, and might win games in the regular season, but both offense and defense are needed to win championships.

Chapter 8
Strategic Implications of the Findings in this Book

8.1 The "So What?" Question

During my Ph.D. studies, I had grand ideas of changing the world and conducting research that would be used to shift paradigms. In the first 2 years of the Ph.D. program I attended at Saint Louis University, I investigated so many questions, crunched so many numbers, and conducted every analysis I knew how to conduct. It seemed as if I was just doing research for the sake of doing research, like a guy who just seeks to investigate relationships between variables for sake of seeing which ones have significant relationships. One day, I was excited about a particular structural equation model I conducted and shared the results with a senior professor. I thought the professor would share my excitement, but instead he said two words to me that would forever change how I approached analytics and research. I remember he removed his glasses, set them on his desk, and he blurted out with a smug look on his face, "so what?". I was confused for a few seconds and silence began to fill the room. The uncomfortable silence made me wonder what in the world he was trying to get out of me. Finally, after I chuckled awkwardly, the esteemed professor explained to me that analytics and research mean nothing unless the researcher can actually express what people in the world can do with the findings of the research. Oh, you found that consumers in Madagascar identify with global brands more than local brands...so what? Oh, you found that Tom Brady is the most efficient athlete of his time...so what? Oh, you found that Lionel Messi is better than Cristiano Ronaldo...so what? Now, after every analysis I conduct, I ask the "so what?" question. So, this chapter answers the "so what" question for the findings of the previous 7 chapters of this book.

To answer the "so what" question, here, I present strategic implications for the findings of each chapter by addressing the issue of how the results of the analyses conducted in each chapter can help sports marketers, sports media members, team general managers, and coaches succeed in their respective roles in the sports world.

A. A. Randrianasolo, *Triple Double*, SpringerBriefs in Statistics, https://doi.org/10.1007/978-3-030-79032-5_8

Also, these results may inspire future researchers in sports analytics to dig further into each topic and hopefully uncover deeper understandings of the inner workings of the variables examined in this book. The following paragraphs discuss such implications.

8.2   Chapter 1: Oh, You Found a New MVP Value Score... So What?

In 2008, scholar Stephen Kershnar published a paper in the *Journal of Social Philosophy*, which stated that MVP should be based on how well a player contributes to his own team (Kershnar, 2008). In other words, this paper states that player value is dependent on how well a player's team performs when the player is on the court versus when the player is not on the court. Three years later, Kershnar refuted his own work in a paper published in the *Journal of the Philosophy of Sport* and states that his previous theory does not actually measure player value (Kershnar, 2011). In fact, Kershnar holds that the MVP problem remains unsolved.

I argue here that the value score developed in Chap. 1 provides a possible solution to the MVP problem. The value score offered in Chap. 1 examines player value with respect to the variables that contribute to winning games in the league rather than just how they impact their own teams. It would be unfair to simply examine the impact of a player on his team since some players have better teams than others, and, therefore, it would give an unfair advantage to players who have worse teams. For example, in 2017, Russell Westbrook won the MVP award and undoubtedly impacted his team more than players such as Stephen Curry because there were no other superstars on Russell's team. This does not necessarily mean that he was more valuable in the league than Stephen Curry. If there were team MVPs awarded, then Russell might win for the Oklahoma City Thunder that year, but the award is for the NBA MVP.

Furthermore, the value score in this chapter accounts for the current era (the sample is composed of teams from 2014 to 2019). This is important to note as what contributed to winning (and value) 10 years ago may no longer be applicable to the current era as the game and league changes. Chapter 1 thus provides two strategic implications for general managers, coaches, and other members of the sports world:

- As the game evolves in the NBA, so does what constitute value. Therefore, when assessing player value, general managers should consider the factors that influence value in the era in which the player they are evaluating is playing. In other words, the influence of each factor in the value score may change depending on the years examined.
- The value score shows more than the usual (points, rebounds, and assists) that is usually cited in MVP discussions. When voting for MVP, members of the media and sports world should consider factors such as turnovers and steals when valuing different players.

8.3   Chapter 2: Oh, You Found that Wilt Chamberlain Is the GOAT…So What?

In a recent research article, Mertz et al. (2017) published a paper in which these scholars investigate what factors influence NBA players to be considered in a list of the greatest of all time. They find that points per game, rebounds per game, assists per game, and championships won were all significant predictors for players to be considered on 150 players in NBA history. The analyses of these scholars' research however fails to account for the differences in what constituted greatness in the different eras of the NBA. As mentioned in Chap. 2, players like Bill Russell and Jerry West did not have a 3-point line, and therefore the game was played in a much different manner than games in 2020 when teams could take upwards of 30 3-point shots per game. The findings in Chap. 2 highlight that no matter if we calculate a GOAT score for the entire NBA history or if we divide the history of the NBA into the 3 eras and calculate a GOAT score for each era, Wilt Chamberlain ranks statistically as the GOAT. These findings provide implications to sports media members:

- Journalists and TV sports personalities often argue these days about whether Lebron has surpassed Jordan in the GOAT debate, but why is Wilt Chamberlain never discussed as the GOAT? Someone who scored 30 points per game and nearly 23 rebounds per game in his career should be at least considered in this discussion.
- Other than implications for the GOAT debate, the results of Chap. 2 also imply that perhaps greatness should be measured only in the playoffs as it was shown that there were major differences when GOAT scores were calculated in the regular season as opposed to the playoffs.

8.4   Chapter 3: Oh, You Found that Super-Teams Do Not Actually Perform Better than Non-Super-Teams… So What?

This chapter provided evidence that the "super-team" phenomenon may be a manifestation of the media or the fans. The fact is, it is rare to have a championship-caliber team that does not have at least 2 superstars, from the Kobe and Shaq Lakers to the Big 3 Celtics to the Hampton 5 Warriors. The findings of this chapter imply that NBA fans, journalists, media members, and TV sports personalities should consider that superstars playing on the same team is not a new phenomenon, yet it is the fact that players yield more power in determining their fate that has some personalities calling them weak. Maybe they are not weak. Maybe they just make boss moves now.

8.5   Chapter 4: Oh, You Found that the All-NBA Selection may not Represent the Best Players in the NBA for the Respective Season...So What?

Making an All-NBA team can have significant effects on a player's career in the short and long term. For example, because Ben Simmons made an All-NBA team in 2020, his salary for the 2020–2021 season is worth 28% of the salary cap rather than just the 25% that he would have made if he did not get the All-NBA selection. Conversely, Joel Embiid did not make an All-NBA team in 2020 and needs to make it in 2021 in order to be eligible for a supermax contract (Mathur, 2020). With such large implications on players, the All-NBA selection process should be an objective process within the NBA rather than a voting process where the subjectivity of the voters allows for the diminishing of relevant variables or the embellishment of irrelevant variables in the selection process. There is already an All-Star selection process that can be used for a popularity contest but lets use more objective measures for All-NBA. The findings in Chap. 4 provide evidence that top performing players each year who do not make an All-NBA selection have similar to or even better numbers than players who make the second or third All-NBA team. With this stated, there are three main implications that can be drawn from this chapter:

- The NBA should use an objective measure to select All-NBA teams. In other words, these selections should not be based on votes.
- Team executives, general managers, and coaches should use criteria other than All-NBA selections to compensate and assess the value of players.
- The panel of media members and sportscasters who vote for All-NBA should keep in mind that there exist inherent biases in the All-NBA team selections. Therefore, they should utilize objective performance metrics when making their votes.

8.6   Chapter 5: Oh, You Found that Small Ball Works when the Team Can Shoot Threes...So What?

Recent research in sports analytics finds that team height, that is how tall the players on a team are, is not significantly related to winning games in the regular season, but may influence winning in the playoffs (i.e., Teramoto and Cross, 2018). Specifically, Teramoto and Cross (2018) find that taller team lineups were associated with winning for teams with lower field goal percentages and higher free throw rates. The results from the analyses in this chapter are in accordance with this finding in the sense that for smaller teams or for small ball to work, teams have to have two components: assists and 3-point shooting. In other words, small ball must have effective passing and effective 3-point shooting.

In his impactful work on investigating the importance of assists in contributing to wins, Melnick (2001) states that *how* a basketball team scores points is more important than the number of points it totals at the end of the game. The findings in this chapter effectively amplify this quote. Small ball teams are effective with assisted 3-pointers, not just 3-pointers in general. So, what does this mean for general managers and coaches? Well, here are two strategic implications for this finding:

- First, when constructing small ball lineups, make sure to have effective 3-point shooting. Shorter teams need to be significantly better 3-point shooters than taller teams in order to win.
- Second, those 3-pointers need to be assisted 3-pointers. Small ball is not effective with iso-play 3-pointers where players "go find their shots." 3-pointers are only effective in small ball when coupled with assists.

As discussed in Chap. 5, the Houston Rockets, led by former MVPs James Harden and Russell Westbrook, decided to go all in on the small-ball method in the 2019–2020 season. When they got to the playoffs, they faced the Oklahoma City Thunder in the first round. In that series, the Rockets averaged 21.4 assists per game and shot 36% from beyond the arc, while the Thunder only averaged 17.9 assists and shot 33% from 3-point range. The Rockets won this series with their small ball lineup as they were superior to the Thunder in 3-point shooting and in assists. In the next round of the playoffs, the western conference semifinals, the Rockets faced the Los Angeles Lakers and shot 37% from 3-point range while having 21.2 assists per game. The Lakers on the other hand shot nearly 38% from beyond the arc and had 25.6 assists per game. The Lakers won the series because not only were they bigger, but they outperformed the Rockets in the two categories that small ball needs in order to be effective: 3-point shooting and assists.

8.7   Chapter 6: Oh, You Found that the Clutch Gene Does Not Exist...So What?

Solomonov et al. (2015) conducted a study to test the validity of the notion that certain players perform better under pressure than others. The results of their study showed that although many "clutch" players may put more individual effort in high-pressure situations (as measured by field goal attempts, points scored, foul drawings, and assists), their performance actually is no better than in low-pressure situations, particularly regarding shooting percentage. Similarly, in their empirical analyses investigating if the "clutch" phenomenon existed in the NBA, Wallace, et al. (2013) find that the results of their analyses provide further corroboration of previous results in the literature that the clutch player in the NBA is a myth and does not exist. The results from Chap. 7 in this book are in line with these previous results from sports analytics research and provides two implications for coaches and general managers:

- General managers should not use the clutch factor or clutch statistics as a factor in selecting which players to recruit, trade, or sign.
- Within games, coaches should utilize the best shooter, rather than the most "clutch" shooter to take shots when the team needs to score.

The clutch phenomenon is amazing as it displays how sports fans and even experienced sports analysts can have quite a bit of confirmation bias relating to player performance on the court. Stephen Curry missed a clutch shot in the 2016 and in the 2019 finals therefore he is not clutch. Max Kellerman stated on ESPN's First Take debate show that with the fate of the universe on the line, he would rather have Andre Iguodala than Stephen Curry take a shot. The important thing to remember is that an overall shooting percentage of 50% is considered a good shooting percentage in the NBA, particularly for guards. A 3-point shooter that averages 40% is considered amazing, which means that a great 3-point shooter misses 6 out of every 10 3-point shots that he takes. So, Stephen Curry missing a shot in the finals might not mean that he does not have some clutch gene, it might just mean that he missed a shot that has a higher probability of not going into the basket than into the basket. The most memorable shots however are shots that are considered to be in "clutch" moments, and, therefore, these shots often make or break a player's reputation as either someone with or without the clutch gene. What we must remember however is that analytical investigations show that the best shooters in clutch moments are actually just the best shooters during any other moment of a game.

8.8   Chapter 7: Oh, You Found that both Offense and Defense Matter in Winning Championships… So What?

The 2003–2004 Dallas Mavericks have been stated by many to have one of the best offenses ever. In his Bleacher Report article, Fromal (2014) ranked this team as the best offense of all time. The Mavs that year ended the regular season with a respectable 52–30 record and faced the Sacramento Kings in the first round of the playoffs. As good as their offense was, the Mavs lost the series 1–4 and completed a first round exit in the playoffs. The Mavs lost this series after shooting under 40% in field goals and under 30% from 3-point range, while totaling 159 defensive rebounds. The Kings on the other hand shot 44% field goals, a little above 36% from 3-point range, and totaled 176 rebounds. These three categories are found to be the determining factors in winning championships from Chap. 7. But wait! Could not it be true that the Mavs just played worse in all aspects of their game, and the Kings just outperformed them? Not exactly. Although the Kings had more steals and assists throughout the series, the Mavs did manage to commit less turnovers, get more blocks, and most impressively out-rebound the Kings. The Mavs outrebounded the Kings because they had 102 offensive rebounds in the 5 games of the series while the Kings snatched just 63 offensive boards. The difference in this series is that the Kings grabbed more defensive rebounds and shot the ball better. Chapter 7 supports this notion that on the offensive end, shooting contributes to winning championships, while on the defensive end, it is defensive rebounds that matter. These findings produce the following implications:

- When constructing teams, general managers and coaches should consider shooting in general, but also pay attention to 3-point shooting.
- Coaches should emphasize the importance of defensive rebounds over offensive rebounds in winning championships.

References

Fromal, A. (2014). *Ranking the NBA's 20 best offenses of all time.* Bleacher Report. Retrieved from https://bleacherreport.com/articles/2185102-ranking-the-nbas-20-best-offenses-of-all-time#:~:text=%20Ranking%20the%20NBA%27s%2020%20Best%20Offenses%20of. Pacers%3A%20106.36.%20League-Average%20Offensive%20Rating%3A%20102.2...%20More%20

Kershnar, S. (2008). Solving the most valuable player problem. *Journal of Social Philosophy, 39*(1), 141–159.

Kershnar, S. (2011). The most-valuable-player problem remains unsolved. *Journal of the Philosophy of Sport, 38*(2), 167–174.

Mathur, A. (2020). *How making All-NBA impacts Ben Simmons, Rudy Gobert, Pascal Siakam.* ClutchPoints. Retrieved from https://clutchpoints.com/nba-news-how-making-all-nba-impacts-ben-simmons-rudy-gobert-pascal-siakam/

Melnick, M. J. (2001). Relationship between team assists and win-loss record in the National Basketball Association. *Perceptual and Motor Skills, 92*(2), 595–602.

Mertz, J. L., Hoover, D., Burke, J. M., Bellar, M., Jones, M. L., Leitzelar, B., & Judge, W. L. (2017). Ranking the greatest NBA players: A sport metrics analysis. *International Journal of Performance Analysis in Sports, 16*(3), 737–759.

Teramoto, M., & Cross, C. L. (2018). Importance of team height to winning games in the National Basketball Association. *International Journal of Sports Science & Coaching, 13*(4), 559–568.

Solomonov, Y., Avugos, S., & Bar-Eli, M. (2015). Do clutch players win the game? Testing the validity of the clutch player's reputation in basketball. *Psychology of Sport and Exercise, 16*, 130–138.

Wallace, S., Caudill, S. B., & Mixon, F. G., Jr. (2013). Homo certus in professional basketball? Empirical evidence from the 2011 NBA playoffs. *Applied Economics Letters, 20*(7), 642–648.

Chapter 9
Debates that Future Work Should Consider

9.1 Why Future Work Is Needed

Thus far in this book, I have attempted to provide empirical resolutions to some of the most pressing debates in the NBA world. From attempting to crown the G.O.A.T. to coming up with a value score to crown MVPs to investigating under which circumstances the small ball method works, among others. However, the debates I present in Chaps. 1–7 do not nearly cover the wealth of debates that NBA fans all over the world discuss on a daily basis. In this chapter, I present debates that future research and books should consider exploring in order to provide us with perspectives on how the NBA that we know and love could be more fair, which would provide benefits to players, fans, and organizations in the league. The following paragraphs discuss the most pressing debates that I believe should be investigated.

9.2 Should the MVP Award Consider Playoff Performance?

As it stands in the NBA, the MVP award is given to the player who performs at the highest level during the regular season. Some analysts and TV personalities have argued however that perhaps the MVP award should include player performance in the playoffs. The argument here is that many NBA players, particularly some of the best players in the league, show their highest value in the playoffs rather than in the regular season. Furthermore, some would argue that the statistics in the regular season may not really reflect the dominance of many players in the league. For example, in the 2016–2017 season, many fans remember that the stars on the Golden State Warriors sat out most of the fourth quarter after building large leads within the first three quarters of games against less dominant teams, and so their per game statistics may not really reflect their value or dominance. In other words, if the

© The Author(s), under exclusive license to Springer Nature Switzerland AG 2021
A. A. Randrianasolo, *Triple Double*, SpringerBriefs in Statistics,
https://doi.org/10.1007/978-3-030-79032-5_9

splash brothers would have played a few more minutes in the fourth quarters of those games when the Warriors had a 20-point lead, they may have been able to boost their stats to compete for the MVP award. Another argument for including the playoffs in the MVP race is that the regular season makes it easy for players on weaker teams to "stat-pad." This sentiment is reflected in the fact that in the past two decades, the MVP of each season has only won the championship four times in their respective MVP seasons, as shown in Table 9.1.

Those who counter the argument that the MVP award should include the playoffs might argue that including the playoffs might diminish the value of the regular season. In other words, fans pay money to go watch their favorite stars play when the stars come to their towns, so taking away an incentive to play at the highest level might result in games of lesser quality in the regular season. After all, why would superstars like Lebron James and Kevin Durant risk getting injured or fatigue themselves in the regular season when all of the incentives and accolades come from good playoff performance? No matter the perspective, one thing is clear: future research should investigate the intricacies of this debate and provide the sports world with objective analyses on whether the MVP award should include the playoffs.

Table 9.1 MVPs between 2000 and 2020 and whether they won the championship that year

Season	MVP winner	Did the MVP win the championship that year?
2019–2020	Giannis Antetokounmpo	No
2018–2019	Giannis Antetokounmpo	No
2017–2018	James Harden	No
2016–2017	Russell Westbrook	No
2015–2016	Stephen Curry	No
2014–2015	Stephen Curry	Yes
2013–2014	Kevin Durant	No
2012–2013	Lebron James	Yes
2011–2012	Lebron James	Yes
2010–2011	Derrick Rose	No
2009–2010	Lebron James	No
2008–2009	Lebron James	No
2007–2008	Kobe Bryant	No
2006–2007	Dirk Nowitzki	No
2005–2006	Steve Nash	No
2004–2005	Steve Nash	No
2003–2004	Kevin Garnett	No
2002–2003	Tim Duncan	Yes
2001–2002	Tim Duncan	No
2000–2001	Allen Iverson	No

9.3   Should the Finals MVP Always Be on the Winning Team?

In the 2015 NBA Finals, the Golden State Warriors beat the Cleveland Cavaliers to win the championship and Andre Iguodala was named Finals MVP. In the series, Iguodala averaged 16.3 points, 5.8 rebounds, and 4 assists per game. Meanwhile, the person he was guarding for a large portion of the series, Lebron James, averaged 35.8 points, 13.3 rebounds, and 8.8 assists per game (Conway, 2015). Lebron actually doubled Iguodala's stats, with twice as many points, assists, and rebounds per game, yet Iguodala walked away with the Finals MVP trophy. So, is it fair that the Finals MVP trophy always goes to the winners of the Finals? One side would argue that the winners already got the championship trophy, while the Finals MVP should be able to go to the player that performs at the highest level in the finals. The other side might argue that if a player and his team lose the finals, they may not be so valuable after all. Either way, here, I am calling for future work to empirically investigate this issue.

9.4   Is the Hall of Fame Process Fair?

Every year, basketball greats are inducted into the Naismith Memorial Basketball Hall of Fame. However, what gets a player into the hall of fame is largely unclear, as the committee that inducts hall of famers has not been transparent about the process. The lack of transparency irks many fans as they wonder why some of their favorite players have not been enshrined in the hall. For example, some have argued that Chauncey Billups, a former Finals MVP and champion should be in the hall. I mean, Chauncey did lead his Pistons team to take down the Kobe and Shaq Lakers in the NBA Finals. Not many players can say the same.

9.5   Is Load Management a Legitimate Reason for Players to Sit During Games?

On November 6, 2019, the Los Angeles Clippers were set to face the Milwaukee Bucks in what many thought would be a clash between the reigning MVP (Giannis Antetokounmpo of the Bucks) and the reigning NBA Finals MVP (Kawhi Leonard of the Clippers). Sadly, many fans were disappointed as the clash of the two titans did not happen that day due to Kawhi Leonard opting to sit out of the game, citing load management (Boggs, 2019). This brings forth the questions: is this fair and is it worth it for players to sit out during the regular season due to load management? In their paper published in the *Orthopaedic Journal of Sports Medicine,* Belk et al. (2017) find no significant benefits of players resting during the regular season with regard to playoff performance. On the other hand, one could argue that players know their bodies and if they need rest, they should have the right to rest. Either way, load management has been a hot topic of debate in the NBA and future research should explore its benefits and drawbacks.

9.6 Is the 2-Conference System the Best System for the NBA?

In the 2017–2018 NBA regular season, the Denver Nuggets did not make the play-offs despite having a 46–36 (56%) regular season record. However, the Washington Wizards [43–39 (52%)], the Milwaukee Bucks [44–38 (54%)], and the Miami Heat [44–38 (54%)] all made the playoffs because the Eastern conference was not as tough as the Western Conference. Is this really fair? Should there be teams that go to the playoffs in one conference who have worse records than teams in another? Would it benefit the NBA to just go with the top 16 teams in the playoffs or is the 2-conference system better? Future research should empirically examine the pros and cons of this debate.

References

Belk, J. W., Marshall, H. A., McCarty, E. C., & Kraeutler, M. J. (2017). The effect of regular-season rest on playoff performance among players in the National Basketball Association. *Orthopaedic Journal of Sports Medicine, 5*(10), 2325967117729798.

Boggs, J. (2019). *Some NBA fans feel ripped off that Kawhi Leonard sat due to 'load management'.* KXLH. Retrieved from https://www.kxlh.com/news/national/some-nba-fans-filled-ripped-off-that-kawhi-leonard-sat-due-to-load-management

Conway, T. (2015). *NBA finals 2015 MVP: Andre Iguodala's stats, highlights and twitter reaction.* Bleacher Report. Retrieved from https://bleacherreport.com/articles/2498418-nba-finals-2015-mvp-andre-iguodalas-stats-highlights-and-twitter-reaction

Printed in the United States
by Baker & Taylor Publisher Services